ROADSIDE GEOLOGY of MISSISSIPPI

ROADSIDE GEOLOGY

of MISSISSIPPI

STAN GALICKI and
DARREL SCHMITZ

2016
Mountain Press Publishing Company
Missoula, Montana

All photos by Stan Galicki and Darrel Schmitz unless otherwise credited.

Geologic maps for the road guides are based on the 2011 *Geologic Map of Mississippi*, published by the Mississippi Office of Geology.

Library of Congress Cataloging-in-Publication Data

Names: Galicki, Stan, 1958- | Schmitz, Darrel W.
Title: Roadside geology of Mississippi / Stan Galicki and Darrel Schmitz.
Description: Missoula, Montana : Mountain Press, 2016. | Includes
  bibliographical references and index.
Identifiers: LCCN 2016033095 | ISBN 9780878426713 (pbk. : alk. paper)
Subjects: LCSH: Geology—Mississippi—Guidebooks.
Classification: LCC QE129 .G35 2016 | DDC 557.62—dc23
LC record available at https://lccn.loc.gov/2016033095

Printed in the United States by Versa Press, Inc.

P.O. Box 2399 • Missoula, MT 59806 • 406-728-1900
800-234-5308 • info@mtnpress.com
www.mountain-press.com

ACKNOWLEDGMENTS

Geologists made this book possible. Geoscience inquiry into the state's geology began in the late 1800s and continues today. The information included in this roadside guide is a result of all the geological research that has been done in the state. We would most like to thank several individuals who directly helped us get this book to press: Laura Barnhart, Rodney Beasley, Cole Lenz, and Leonard Rawlings, Jr., all of whom spent countless hours navigating thousands of miles of Mississippi roads with a geology map in one hand and a road map in the other while trying to navigate, drive, and integrate the two maps.

We would like to thank Dr. David Dockery for his meticulous documentation of Mississippi geology during his tenure at the Mississippi Office of Geology. His encyclopedic work *Mississippi Geology* is a thorough documentation of the geology throughout the state. His input not only helped us develop the geology presented in this book but also put the geology into a historical perspective relative to human settlement. James Starnes, also with the Mississippi Office of Geology, was always ready to answer questions related to surface geology and surface processes and provided valuable input during editing. The coastal section of this book would not have been complete without the contributions of Dr. Ervin Otvos. His knowledge of coastal geology was critical in our understanding of the Quaternary geologic history of coastal Mississippi.

We would also like to thank our wives, Kristie and Donna, for their continued support throughout this project, their miles and miles behind the wheel, and tolerating endless stops at exposures across the state of Mississippi.

Finally, we would like to thank our editor at Mountain Press Publishing, James Lainsbury. This book is unlike anything we have ever written (or will likely write again). He patiently guided us through the editing of the draft and made significant contributions to maps and figures. We came to trust his judgment and value his input while editing this project. We cannot thank him enough.

PREFACE

Advances in technology continually increase our knowledge base and thus our understanding of the Earth. Much of our insight into Mississippi's geologic history is the result of careful surface mapping, seismic data, and data generated from the drilling of water, oil, and gas wells. As researchers conduct more studies, the intricacies of Mississippi's geology come to light, and ideas once thought to be well established are opened for interpretation. There are topics of Mississippi geology in which researchers do not find agreement. In this book we strove to develop a well-balanced presentation of Mississippi geology and provide input from many perspectives. State geologists are currently reexamining the surface geology across the state and assessing old interpretations. The geologic map of the state will undoubtedly change, but the maps published in this book are as up-to-date as possible.

Our colleagues at the Mississippi Office of Geology reminded us that some of the best geology in the state is well off major highways. Recognizing this, we included a few of those places that are most accessible. Much of Mississippi's geology must be interpreted by changes in the landscape. We pointed out the relationship between topography, geology, and even ecology whenever it seemed relevant to the discussion.

Due to the loosely consolidated nature of most of Mississippi's sediments, they do not resist the forces of erosion nor stand in outcrop very well. Mississippi's high annual rate of precipitation and warm temperatures result in high rates of erosion and the growth of vegetation, both of which tend to obscure natural outcrops. Historically, road cuts provided excellent windows into near-surface geology, but the Mississippi Department of Transportation is quick to grade eroding embankments and seed most exposures. The photos in this book were taken in areas that over the past few decades have withstood the forces that work to destroy them.

We hope this book helps readers understand the complexity of Mississippi's geologic history. Enjoy.

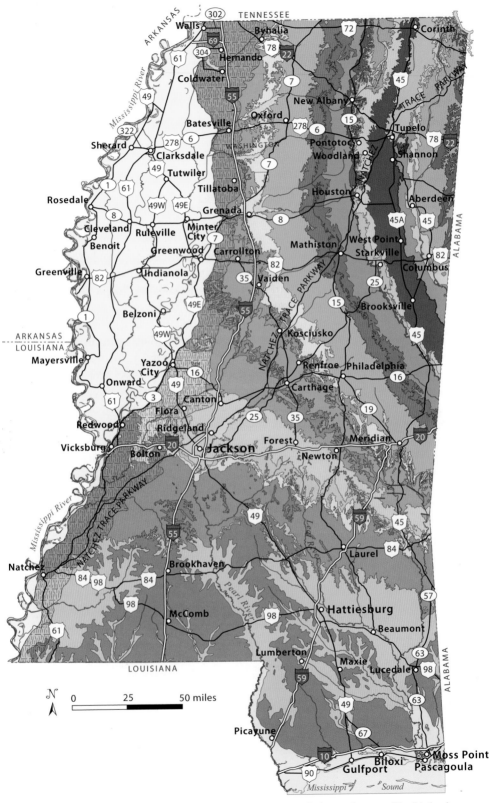

Simplified geologic map (and legend) of Mississippi with the roads covered in this book.

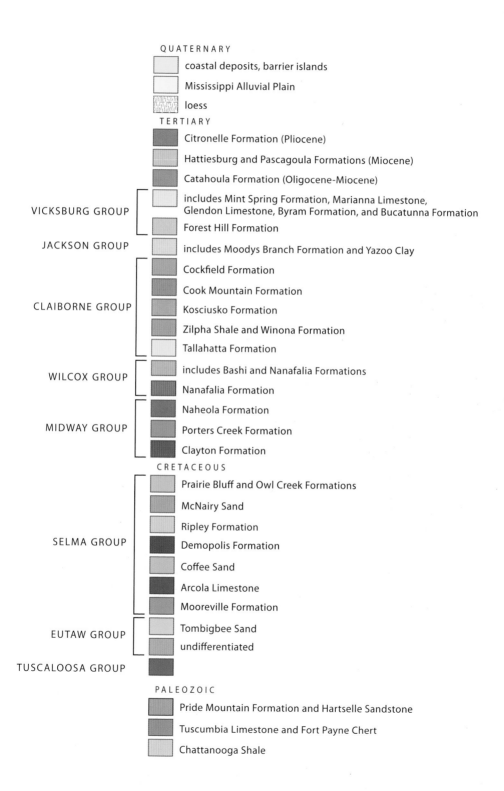

QUATERNARY

coastal deposits, barrier islands

Mississippi Alluvial Plain

loess

TERTIARY

Citronelle Formation (Pliocene)

Hattiesburg and Pascagoula Formations (Miocene)

Catahoula Formation (Oligocene-Miocene)

VICKSBURG GROUP

includes Mint Spring Formation, Marianna Limestone, Glendon Limestone, Byram Formation, and Bucatunna Formation

Forest Hill Formation

JACKSON GROUP

includes Moodys Branch Formation and Yazoo Clay

CLAIBORNE GROUP

Cockfield Formation

Cook Mountain Formation

Kosciusko Formation

Zilpha Shale and Winona Formation

Tallahatta Formation

WILCOX GROUP

includes Bashi and Nanafalia Formations

Nanafalia Formation

MIDWAY GROUP

Naheola Formation

Porters Creek Formation

Clayton Formation

CRETACEOUS

Prairie Bluff and Owl Creek Formations

McNairy Sand

Ripley Formation

SELMA GROUP

Demopolis Formation

Coffee Sand

Arcola Limestone

Mooreville Formation

EUTAW GROUP

Tombigbee Sand

undifferentiated

TUSCALOOSA GROUP

PALEOZOIC

Pride Mountain Formation and Hartselle Sandstone

Tuscumbia Limestone and Fort Payne Chert

Chattanooga Shale

CONTENTS

GEOLOGIC TIMESCALE

EON	ERA	PERIOD			AGE (millions of years)	GEOLOGIC EVENTS IN MISSISSIPPI
				EPOCH		
PHANEROZOIC	CENOZOIC	QUATERNARY		HOLOCENE		Humans become agents of geologic change. Mississippi River assumes present configuration. Sea level rises and current barrier islands form.
					0.01	
				PLEISTOCENE		Loess deposits form on eastern side of Mississippi River Alluvial Plain. Glacial stages result in lower sea level and stream down-cutting; Mississippi River down-cutting enables the creation of Mississippi River Valley Alluvial Aquifer.
					2.6	
		TERTIARY	NEOGENE	PLIOCENE		Mississippi coastline moves southward out of the Mississippi Embayment.
					5	
				MIOCENE		
					23	
			PALEOGENE	OLIGOCENE		Vicksburg Group records the last major marine transgression in the Mississippi Embayment.
					34	
				EOCENE		Multiple transgressions and regressions in the Mississippi Embayment; Yazoo Clay deposited during a transgression. Whales thrive in coastal water. Lignite deposition continues in coastal areas.
					56	
				PALEOCENE		Widespread deltas; wetlands result in lignite deposits.
					66	
	MESOZOIC	CRETACEOUS				Chalk deposition widespread in the Mississippi Embayment. Jackson and Midnight volcanoes active. Continued clastic sedimentation in Mississippi Interior Salt Basin and Gulf of Mexico.
					145	
		JURASSIC				Widespread deposition of Louann Salt in Mississippi Interior Salt Basin and Gulf of Mexico, followed by carbonates, and then clastic sediment.
					201	
		TRIASSIC				Pangaea begins to rift apart; Mississippi Interior Salt Basin and Gulf of Mexico open.
					252	
	PALEOZOIC	PERMIAN				No rock record in Mississippi. Ancestral Appalachians and Ouachitas weather and erode.
					299	
		PENNSYLVANIAN				Ancestral Ouachita Mountains form as Gondwana (Europe and Africa) approaches from the southwest. Black Warrior Basin fills with sand and coal deposits.
					323	
		MISSISSIPPIAN				Ancestral Appalachians rise to the east and southeast as Gondwana (Europe and Africa) approaches. The Black Warrior Basin receives sand and mud from northern and eastern sources.
					359	
		DEVONIAN				Iapetus closes to the east; sedimentation shifts from carbonate to silica-rich clastics.
					419	
		SILURIAN				Iapetus widens to its peak and then begins to close.
					444	
		ORDOVICIAN				An extensive carbonate bank forms across southern Mississippi.
					485	
		CAMBRIAN				
					541	
PRECAMBRIAN						The proto-Atlantic Ocean, or Iapetus, forms. The short-lived supercontinent Pannotia forms, rifting apart between 600 and 540 million years ago along the Alabama-Oklahoma Transform Fault. Rodinia rifts apart 800 to 750 million years ago, creating the Mississippi Valley Graben. Earth forms about 4,600 million years ago.

Geologic timescale. (Modified from Geological Society of America 2012.)

THE BILLIONS YEARS WAR

It is inconceivable to most people that for the past 4.6 billion years a war has raged on Earth between two great forces. Solar radiation from the sun drives our weather patterns, unleashing wind, rain, and ice to weather and destroy every rock that has dared rise out of the sea. Not to be outdone, the restless Earth fights back by forming mountain ranges and elevated terrain. The Earth's ability to fight back originates from internal heat emanating from its core and generated by the radioactive decay of unstable elements beneath the crust. Most of the battles take place at such an incredibly slow rate that humans are unaware that they are even occurring. Occasionally, humans are caught in the violent fury delivered by the sun, as evidenced by hurricanes, tornadoes, and floods. Not to be outdone, the Earth sends forth earthquakes, volcanic eruptions, and tsunamis. Although these natural hazards are often destructive from a human perspective, they are evidence that the Earth is very active and fighting back. When the Earth's thermal energy has dissipated, these events will not occur; the Earth will not be able to fight back, and all land will erode to the sea. Humans have proven to be a formidable species, but we are helpless to control the forces that shape the Earth around us. Geology helps us understand where previous earthly battles occurred, where they are likely to occur, and how best to avoid getting caught in the disasters that will occur again.

Geologists, like all professionals, work with ideas unique to their field. Two issues that geologists always have in the back of their mind are geologic time and the rock record. Time is everything to a geologist; give a process enough time, and mud on the ocean floor can become rock and be moved from the abyssal depths of the ocean to the top of the highest peak on Earth. The average global human life expectancy is around seventy years, so it is not surprising that changes that take place over thousands, millions, or even billions of years are hard to conceive. Some geologic events are catastrophic and instantaneous, such as earthquakes and tsunamis, but most geologic events take place at very slow rates. Regardless, most events leave behind evidence that becomes part of the rock record, and it's the rock record that geologists use to interpret Earth's history.

Assembling the 4.6-billion-year puzzle that constitutes Earth's history is complicated because some of the pieces are gone; there are gaps in the rock record. These gaps occur for a variety of reasons. Rocks exposed to the elements weather and disappear, taking with them information about what Earth's surface looked like when the rocks were forming; others are deeply buried and

1

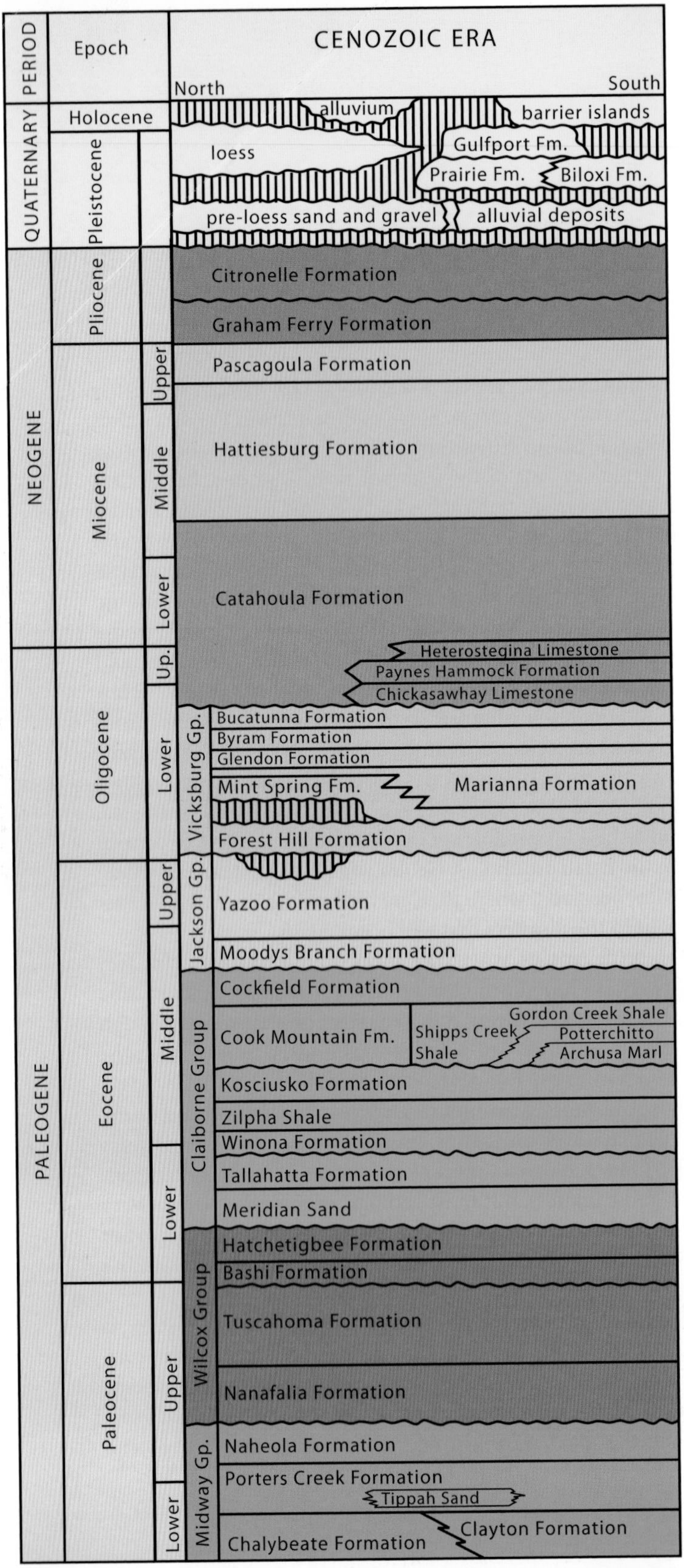

CENOZOIC ERA

PERIOD	Epoch			Stratigraphic units

North ... South

QUATERNARY

Holocene — alluvium, barrier islands
Pleistocene — loess, Gulfport Fm., Prairie Fm., Biloxi Fm., pre-loess sand and gravel, alluvial deposits

NEOGENE

Pliocene
- Citronelle Formation
- Graham Ferry Formation

Miocene
- Upper: Pascagoula Formation
- Middle: Hattiesburg Formation
- Lower: Catahoula Formation

PALEOGENE

Oligocene
- Up.: Heterostegina Limestone, Paynes Hammock Formation, Chickasawhay Limestone
- Lower (Vicksburg Gp.): Bucatunna Formation, Byram Formation, Glendon Formation, Mint Spring Fm. / Marianna Formation, Forest Hill Formation

Eocene
- Jackson Gp.
  - Upper: Yazoo Formation, Moodys Branch Formation
- Claiborne Group
  - Middle: Cockfield Formation; Cook Mountain Fm. — Shipps Creek Shale / Gordon Creek Shale, Potterchitto, Archusa Marl; Kosciusko Formation; Zilpha Shale; Winona Formation
  - Lower: Tallahatta Formation, Meridian Sand

Paleocene
- Wilcox Group
  - Upper: Hatchetigbee Formation, Bashi Formation, Tuscahoma Formation, Nanafalia Formation
- Midway Gp.
  - Naheola Formation
  - Porters Creek Formation — Tippah Sand
  - Lower: Chalybeate Formation — Clayton Formation

Stratigraphic columns for Cenozoic-, Mesozoic-, and Paleozoic-age geologic units in Mississippi. Wavy horizontal lines are unconformity surfaces. The vertical hash marks between them represent extended periods of erosion. When two units are adjacent to one another and separated by a jagged vertical line, they were deposited at the same time but in different environments. Upper, middle, and lower designations refer to the position of rocks in the period during which they were deposited. Not all of the units are exposed at the surface; the Cenozoic units are the rocks most commonly seen in Mississippi. (Modified after Dockery 2008a, 2008b, and 2008c.)

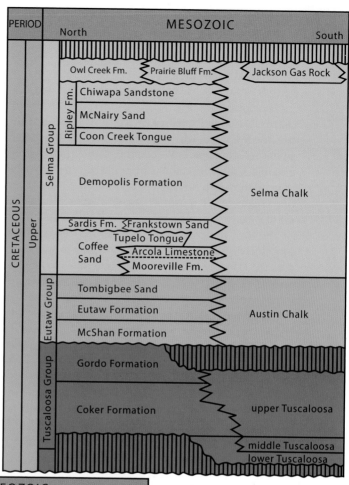

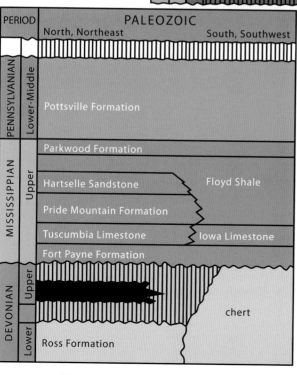

far removed from surface processes for millions of years; and sometimes long periods of time elapse when no sediments are deposited at the surface, meaning geologists have no evidence to interpret. Missing time in the rock record is called a hiatus and is represented by a surface called an *unconformity*. The surface you live upon, where Earth's geologic history is being eroded, will become an unconformity far in the future. Go out into the ocean, however, and you will find that sediment removed from the continent by wind, rivers, and ice is settling on the ocean floor; there Earth history is being recorded. Throughout all of geologic time, a record was made at some place on Earth's surface; however, for the reasons noted above, we may never be able to investigate that record.

Traces of some of Earth's earliest rocks are exposed in the heart of many continents or are revealed at great depths in well borings. Pioneering geologists used these scattered pieces of the rock record as well as younger ones to assemble the geologic timescale, which tells the story of Earth from its genesis. Generations of geologists have developed and refined the timescale, a process that continues today as more information is uncovered around the world.

The principles that led to its creation were first being developed in the late 1600s. However, one of the first active contributors to the timescale was William Smith, a surveyor and engineer in England. Conclusions he made about the relationship between fossils and sedimentary rocks while building canals in England resulted in the first geologic map, made in 1815. As more people recorded observations regarding changes in fossils that occurred over time, the geologic timescale evolved. Geologists now have tools other than fossil assemblages to date rocks, the most significant being radiometric dating (see "Dating Rocks").

The greatest divisions of time on the timescale are called *eons*, of which there are four: Hadean, Archean, Proterozoic, and Phanerozoic. The first three are often grouped together and called the Precambrian. The Hadean, Archean, and Proterozoic comprise 88 percent of all geologic time. *Proterozoic* means "first," whereas *Phanerozoic* means "visible life." These titles are in reference to the change in life-forms from largely microscopic, unicellular organisms to multicellular, higher-order forms.

The Phanerozoic is divided into three eras: Paleozoic, Mesozoic, and Cenozoic, their names referring to "early," "middle," and "recent" life respectively. Although recognized extinction events are evident in many places in the timescale, two major extinction events mark the boundaries between these three eras. The Permian extinction, the greatest to have occurred on Earth, closed out the Paleozoic. The Cretaceous extinction, which eliminated the dinosaurs, closed out the Mesozoic. Each era is divided into shorter intervals called periods, which are in turn divided into epochs. We live in the Holocene Epoch of the Quaternary Period. Although nothing formal has been recognized, some scientists are suggesting that a new epoch be created called the Anthropocene because of the obvious impact humans have had on Earth's atmosphere and landscape. Questions remain about when exactly human impact became significant. Was it the dawn of agriculture, the start of the industrial revolution, or the detonation of the first nuclear weapon? The discussion continues.

Geologists name rocks for the area where they first encounter them. For example, the infamous Yazoo Clay, responsible for so much foundation damage across central Mississippi, was first studied along a bluff on the Yazoo River in Yazoo County. It was named in 1915.

Formation is the basic term for a geologic unit. Formations are readily identifiable, distinct, and can be mapped over large areas at the surface or in the subsurface. Several formations that are related in time or by depositional environment may be referred to as a *group*. Rocks that are part of a formation or group may have been deposited over millions of years. Once an age is established for a geologic formation using radiometric dating or fossil identification, geologists assign that rock unit to the interval of time in which it was deposited. The Yazoo Clay, for example, was deposited toward the end of the Eocene Epoch, which ranged from 56 to 34 million years ago.

PLATE TECTONICS

Evidence that the Earth's surface is losing the battle is all around us. Degradation of the landscape via weathering and erosion is more apparent than the building of mountains. A mudslide or rock fall; a dust storm; a floodwater-swollen river darkened by huge amounts of suspended clay, silt, and sand; and even the load of sediment deposited at the end of a glacier are all examples of how sediment is transported after rock has been broken down by weathering processes. Wind, water, ice, and gravity are the agents that move mountains from the land to the sea.

How exactly landscapes are built, however, is less apparent. Volcanic eruptions and earthquakes are formidable natural hazards. Although these geologically instantaneous events are viewed as destructive from a human perspective, they are signs that the Earth is still full of energy and fighting back.

As the battle rages on, continents and islands continually rise out of the sea only to be broken down and returned to the sea. Geologists refer to this cycle of destruction and creation as the *rock cycle*. The cycling of Earth materials also provides that mineral deposits are emplaced, oil and natural gas deposits accumulate, and fertile soils are deposited, all of which are natural resources critical to human evolution and cultural development.

Geologists at the dawn of the twentieth century thought they understood the intricacies and workings of the Earth. It is not surprising that when a new theory called *continental drift* was first presented, using ideas developed in 1596, it was scoffed at. In the early 1900s, meteorologist Alfred Wegener hypothesized that it was not coincidence that the eastern margin of South America and the western margin of Africa would fit together like puzzle pieces if the Atlantic Ocean wasn't present. He could not ignore the landscape trends that were created when he reorganized the continents into a single landmass he called Pangaea: glaciated terrains and mountain ranges lined up, fossil groupings and rock strata became contiguous. The evidence that the continents must have drifted apart over long spans of time was there, but Wegener lived in a time that did not have the technology to substantiate these claims. Wegener's continental drift faded from the spotlight of discussion as governments and scientists became preoccupied with finding and utilizing natural resources and fighting two world wars. The true nature of the rock cycle was not apparent until the middle of the twentieth century.

A series of technological developments in the late 1950s, some a direct result of wartime research, led to a whole new understanding of what the interior of the Earth looks like and how it behaves. These advancements in scientific understanding vindicated Alfred Wegener's theory.

The radius of the Earth is nearly 4,000 miles. The Russians, drilling on the Kola Peninsula, have drilled the deepest hole on Earth, but it is only just over 7.5 miles deep. How then do geologists know what lies within the Earth? Some geoscientists study how rocks behave using high-pressure and high-temperature instruments to mimic the extreme conditions of the deep Earth. Others study the chemistry and properties of rocks produced deep within the Earth, such as kimberlite deposits (where diamonds form), or lavas from volcanic terrains, such as those in Hawaii. Most details of the Earth's interior structure, however, come from seismic investigations. Geophysicists study the behavior of sound waves, such as those generated by earthquakes, as they travel through the Earth. Their work is similar to how a physician uses CAT scans and MRIs to study the human body.

From decades of gathered data, geoscientists have determined that the Earth's interior is composed of four primary shells known as the *inner core, outer core, mantle,* and *crust.* Each shell has a slightly different chemistry and physical properties. The inner and outer cores are composed of iron- and nickel-rich rock, but the inner core is solid, whereas the outer core is liquid. The mantle, which is more plastic than liquid, occupies the greatest volume and is composed of iron, magnesium, silicon, and oxygen. The crust is the relatively thin skin that covers the planet and is enriched in silicon and oxygen. This basic knowledge provided the foundation from which geologists more fully developed Wegener's theory.

In the late 1950s, sonar initially developed for submarine navigation revealed a 50,000-mile-long chain of mountains (known as a *mid-ocean ridge*) and depressions (known as trenches) nearly 7 miles beneath the ocean surface. Deep-sea drilling projects led to the discovery that the oceanic crust is not more than 180 million years old—much younger than most of the rock composing the continents, and that it is composed primarily of volcanic rock. Even more intriguing, the oceanic crust is youngest at the summit of the submarine mountain chain and gets progressively older toward the margins of the continents.

These discoveries were significant, but another line of research indicated that the Earth's magnetic poles periodically reverse. Today, magnetic north is toward the North Pole, but this has not always been the case. When geophysicists recorded the magnetic signature of the rock on the ocean floor across the summit of the submarine mountain chain, they discovered that the signature (whether it was positive or negative) of the magnetic reversals and the ages of the rock were identical on either side of the ridge. This evidence pointed to a dynamic ocean floor, and it indicated that oceans could indeed open and continents could move—as Wegener had theorized.

As technological innovations and research were integrated, it appeared that continental drift was alive. The continental drift theory was renamed *plate tectonic* theory in 1973. The plate tectonic theory states that Earth's outermost layer, known as the *lithosphere*, is rigid and broken into large or small fragments called *plates*. The plates, which create a jigsaw pattern of boundaries, can be oceanic or continental. Oceanic plates are mostly dense igneous rock, such as basalt and peridotite, that is rich in iron, magnesium, and calcium. Continental plates are mostly less dense igneous rock, such as granite, that is rich in silicon, potassium, sodium, and aluminum. The oceanic lithosphere is about 10 percent more dense than the continental lithosphere. That may not sound like much, but it's enough to cause oceanic plates to sink deeper into the underlying asthenosphere, creating ocean basins, and to be subducted (forced to go beneath) and overridden by the continental crust where two plates meet.

There are approximately fifteen major tectonic plates and many more microplates. Lithospheric plates float on a layer of rock known as the *asthenosphere*. The increased pressure and temperature conditions in the asthenosphere are just right to cause partial melting in the rock, enabling crystals within the rock to move against each other. The asthenosphere rock is far from a molten state, but it does have the ability to flow. Geologists often describe the behavior of the asthenosphere as being "plastic."

Heat energy within the Earth causes thermal convection in the asthenosphere, which causes the lithospheric plates to move about the surface. It's helpful to think of a pot of spaghetti sauce to understand convection. As the stove burner heats the bottom of the pot, the warmed sauce rises to the top, where it cools and sinks back to the bottom of the pan to be reheated again. Asthenosphere rock flows in much the same way. Hotter rock closer to Earth's hot core rises until it reaches the cooler lithosphere. The hot rock then moves horizontally along the base of the lithosphere, cooling before sinking back toward the core. As it moves along the base of the lithosphere, it pushes, pulls, and even drags the lithosphere along with it—voila! Plate movement.

Because the Earth does not increase or decrease in size, plates would not be able to move unless there are places where the lithosphere is created or destroyed. Lithosphere is created at mid-ocean ridges, where magma rises to the surface and cools. A mid-ocean ridge is one of the three types of plate boundaries. At this type of boundary plates slowly separate and move away from each other. Remember that intriguing evidence about the rock getting younger the farther away the researchers were from that submarine mountain chain? This process is why.

Lithosphere is destroyed at what are called *subduction zones*, where an oceanic plate descends beneath a continental plate and dives into the asthenosphere, where it melts. As the plate melts, it generates magma that travels up through the overriding plate, creating a line of volcanoes. Western North America occurs along a subduction zone boundary, and the Cascades are the resulting volcanoes. The third type of boundary is called a *transform boundary*, where plates slide horizontally past one another. The San Andreas Fault is a famous boundary of this type.

As plates move, ocean basins open and close, mountains are pushed upward, and continents are torn apart. And sometimes, as Wegener hypothesized, all of Earth's continents are jammed together to create one giant landmass, or supercontinent. Pangaea was one of these. Current studies indicate that some plates can move up to 5 inches per year. Each earthquake or volcanic eruption—most frequently located at a plate boundary—signals that the Earth is active and changing and fighting back against the climatic forces that work to tear it down.

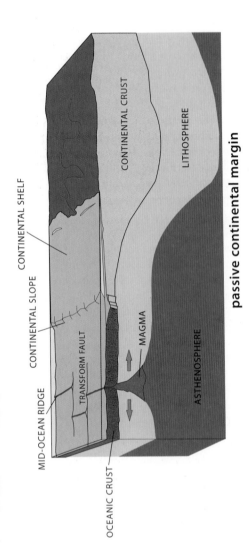

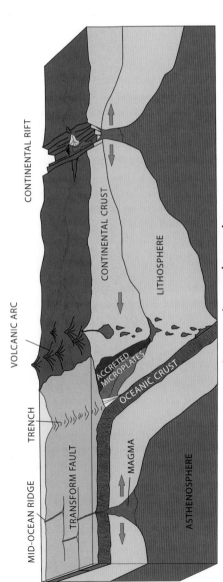

Active and passive continental margins with plate tectonic boundaries and internal features of the Earth. Arrows indicate direction of movement.

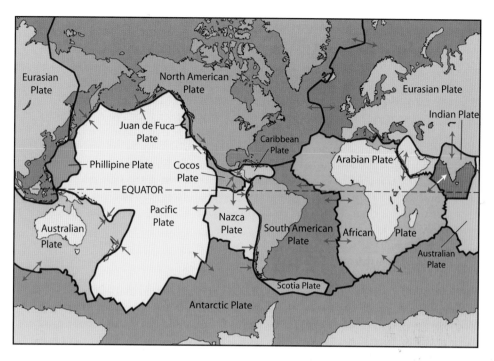

Earth's lithospheric plates and boundaries. Arrows indicate the direction of movement.

Plate tectonics explains why some areas are petroleum rich, others are coal rich, and still others have reserves of diamonds or iron or even salt. It explains why some areas are prone to earthquakes and volcanic eruptions and others are not. Plate tectonics theory was the user's guide to Earth that geologists had been searching for. With a new understanding of how Earth operated, geologists began rewriting its history in the early 1970s, a process that continues today.

THE ROCK CYCLE

There are three major rock types: igneous, sedimentary, and metamorphic. Igneous rocks are created by the cooling of molten rock, which is called *magma* when located beneath the surface and *lava* when exposed at the surface. Sedimentary rocks are created by the compaction and cementation of sediment derived from preexisting rocks (sandstone, siltstone, claystone, and shale), the accumulation of plant material (lignite and coal), or by the chemical precipitation of mineral salts (limestone, salt). Metamorphic rocks form when a preexisting rock is subjected to increased heat, pressure, or chemically active fluids. Given adequate time and the proper processes, any rock type can be broken down or changed to another rock type. This continual cycling of Earth materials is known as the rock cycle. The previously mentioned rock record is the result of the rock cycle. Plate tectonics activity provides the mechanisms that permit the rock cycle to take place.

SEDIMENTARY ROCKS

All the rocks and sediment exposed at the surface in Mississippi are classified as sedimentary. The majority of the rock and sediment is rich in either silica or carbonate. *Silica rich* means that the material is rich in silicon and oxygen. The most common minerals in Mississippi's silica-rich rocks are quartz and members of the clay family. Carbonate rocks, such as limestone and chalk, are primarily composed of calcium carbonate, also known as the mineral calcite. These rocks form as carbonate minerals precipitate from seawater and accumulate on the seafloor. They often are composed of the calcareous remains (shells) of sea life.

Sediment particles are grouped according to their average diameters. In increasing particle size geologists refer to clay, silt, sand, pebbles, cobbles, and boulders. Be aware that the word *clay* can refer to both size and mineralogy. An approximate reference size for sand particles is between that of the period at the end of this sentence and a lowercase *o*. Silt and clay, together called mud, are finer grained, whereas pebbles, cobbles, and boulders, collectively known as gravel, are larger.

Energy is an important factor in determining where a particle ends up. Whether transported by wind or water, it is not difficult to see that it takes much more energy to move a boulder than a grain of silt or sand. When a grain becomes too heavy to be moved, it comes to rest; geologists describe this as deposition. Once deposited, grains are buried. As water flows through the spaces, or pores, between the grains, chemical reactions take place that cause lithification. Lithification is the process that changes sediment to rock and is a combination of compaction and cementation. Compaction forces out all excess water in the pores and permits minerals to grow between the grains, cementing them together and forming rock. Rock strength depends on many factors. Some rocks are extremely well cemented and present a challenge to even the largest of hammers, whereas others crumble in the palm of your hand. Most lithification takes place at considerable depth beneath the surface.

One of the most dynamic aspects of the sedimentary rock formation process is how thousands of feet of sediment can accumulate in one place. How are sediments buried? As more and more sediment is delivered to the site of deposition, the weight of all that material gradually causes the lithosphere to subside. As more sediment is delivered, more subsidence takes place. Mississippi has been situated at the continental margin of North America for millions of years. As we will see when we investigate the subsurface of Mississippi, tens of thousands of feet of sediment accumulated beneath the state.

The area where sediment comes to rest is the depositional environment. Depositional environments can be terrestrial, marine, or transitional between the two. Examples include beaches, stream channels, deserts, reefs, and the deep ocean floor. Sedimentary rocks are deposited horizontally in layers known as beds. Every environment has characteristic bedding that holds clues about its formation. Each depositional environment also has unique physical character-istics that control factors such as grain size, grain sorting, grain shape, fossil content, and the size and shape of the rock unit. For example, beds that are uniformly composed of marine clay were likely deposited on the floor of a deep ocean, clay being a small sediment that can be carried long distances from a shoreline before settling out.

DATING ROCKS

The timing of events is crucial in understanding Earth's history. Earth materials can be dated in a relative sense by describing the order in which events occurred. For example, if a sedimentary rock layer has been intruded by magma, then the sedimentary rock unit must be older than the magma that intruded it. Relative location of layers can also be used. For example, the sedimentary layer on top of another one must have been deposited after the one beneath it; thus, the layer on top is younger. Fossils contained in sedimentary layers also help date rocks if the age of the fossils can be determined. Early geologists worked out much of Earth's history by placing rock units in a relative sequence using careful obser-vation of rock unit boundaries and fossil evidence.

When geologists were piecing together this history, they faced one key dilemma: no one knew exactly how old the Earth was. The discovery of radio-activity in 1896 paved the way for absolute dating techniques. As early as 1907, scientists were using naturally occurring radioisotopes to determine the age of rocks in numeric terms. Radioactive isotopes decay naturally at a measurable rate (half-life) to produce daughter isotopes of different compositions. For a given rock sample containing such isotopes, scientists use the rate of decay and the ratio between the amount of parent and daughter isotopes to calculate its age. Both relative and absolute dating techniques are crucial in interpreting the rock record.

The texture and chemistry (mineralogy) of igneous, sedimentary, and met-amorphic rocks can provide unique clues for interpreting Earth history, but interpretation becomes a bit more interesting when multiple rock types exist together in the same place. The mineralogy and texture of grains or fossils in a sedimentary rock can provide clues to its environment of deposition, rela-tive age, and sometimes even an absolute age in years. Igneous rocks can help researchers understand the tectonic history of an area and provide excellent age information, but they cannot provide information about environmental processes in an area. Metamorphic rocks indicate that the pressure and tem-perature changed at some time after the earliest rocks in an area formed. Dur-ing most metamorphic events, in which rocks are subjected to intense heat and pressure, a rock's textures and mineralogy are altered, and its absolute age is reset. The absolute ages obtained from metamorphic rocks reflect the age of the event that caused the change, not the age of the original rock. Reconstructing Earth history requires investigating all aspects of rock in a region, including physical properties, chemical composition, fossil content, and age. One final aspect of putting together the geologic history of an area relates to its structural geology.

STRUCTURAL GEOLOGY

Structural geology is the study of rocks that are bent, folded, and even broken. These changes to the face of the Earth, which are due to the movement of tectonic plates and the cycling of Earth's materials via the rock cycle, normally occur on a large scale over long periods of time. Catastrophic changes that take place in a matter of seconds can also result in folded and broken rocks. The stress generated by plate movement at major boundaries is transmitted through the lithosphere and causes rocks to warp upward into anticlines and domes or downward into synclines or basins. Anticlines and synclines are linear features in the crust, whereas domes and basins are circular.

When rocks break due to tectonic stress, faults form. Faults are breaks in rocks along which some movement has taken place. Each fault records the types of stress that affected the local area historically or are affecting it at the present time. There are three types of faults: normal, reverse, and strike-slip. The surface along which the movement occurs is called the *fault plane*. In normal and reverse faults the fault plane may be vertical, but it is typically at some other angle, and the primary motion along the fault is vertical. The areas on either side of the fault plane are called blocks. The *hanging wall* block always rests on the *footwall* block. In a normal fault the hanging wall block drops down relative to the footwall block. In a reverse fault, the hanging wall block moves up relative to the footwall block. When the angle of the fault plane is very low (less than twenty degrees), the reverse fault is called a *thrust fault*. Large sheets of continental rock can move horizontally hundreds of miles over other rock along thrust faults.

Normal faults occur where tensional forces have stretched rocks, such as at rift valleys and mid-ocean ridges; reverse faults are caused by compression, such as that which occurs at subduction zones and where continents collide. The primary movement along strike-slip faults is horizontal. It is not difficult to imagine a fence line or a road sliced by a fault and displaced in opposite directions along the fault. A transform fault is a special type of strike-slip fault that connects mid-ocean ridge segments. The San Andreas Fault is a special feature. It's a transform fault that is the boundary between the North American

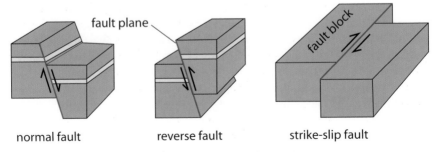

Fault types. Arrows denote direction of movement.

and Pacific plates. Coastal cities such as Los Angles, Santa Barbara, and Monterey are slowly moving to the northwest, toward San Francisco, along the fault as the two plates grind past one another.

Adding an understanding of structural geology to the overall picture enables a geologist to fully interpret the geologic history of an area.

GEOLOGIC HISTORY OF MISSISSIPPI

When investigating only the surface geology of Mississippi, it does not take long to realize that most of Earth's history is not represented. The Devonian- and Mississippian-age outcrops in the northeastern corner of Mississippi—the oldest in the state—represent less than 1 percent of the outcrops in the state. All of the remaining rock outcrops were deposited from the Cretaceous Period on, beginning 145 million years ago.

In order to reconstruct the portion of Mississippi's geologic history that occurred before the Devonian Period (419 million years ago), we must look below the surface and beyond the state's borders—to regional and even global scales—to gain perspective on what happened here. It is at this scale that we can see scars from Mississippi having been caught in the center of the geologic battlefield. All of Mississippi's Precambrian and most of its Paleozoic rocks are buried beneath central Mississippi in the Black Warrior Basin, Ouachita Tectonic Belt, and the Mississippi Interior Salt Basin. Although the rocks are not available at the surface for direct interpretation, geologists have taken advantage of information obtained from the drilling of oil and gas wells and seismic surveys in the southeastern states to piece together the geologic history of Mississippi.

Unlike much of western North America, which was added to the continent through a process known as accretion, the area that would become Mississippi has occupied a coastal position since roughly 600 million years ago and has experienced little accretion. The buried rocks tell the tale of being torn apart twice during the creation of ocean basins and squeezed back together during the assembly of two supercontinents.

Mississippi's story begins with the breakup of Rodinia (roughly between 800 and 750 million years ago), the first of at least three recognized supercontinents in Earth's past. The rifting of Rodinia did not result in what we recognize as North America, but deep-seated faults and volcanic rocks below the surface in Mississippi document a failed rift known as the Mississippi Valley Graben, which is related to Rodinia's demise. The Mississippi Valley Graben will figure prominently into the geologic history of Mississippi much, much later. Over the 450 million years that followed Rodinia's breakup, the Earth's continents came together to form two more supercontinents: Pannotia and Pangaea. During this time, the area that would become Mississippi was situated south of the equator.

Pannotia was a short-lived supercontinent that began to break up, or rift, 600 to 540 million years ago, in late Precambrian time. Land plants had not evolved, so the terrain was quite barren. Pannotia broke apart along a series

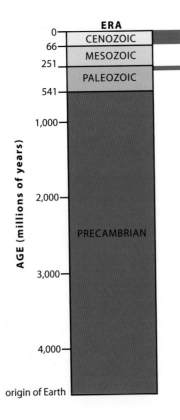

Geologic timescale with the ages of rocks exposed in outcrop in Mississippi (red). Most of the rocks exposed in Mississippi are relatively young. Mississippi's geology is seemingly simple at the surface and represents only 10 percent of Earth's history.

of rifts (divergent plate boundaries) and transform boundaries. Large structural features known as grabens have been mapped in Mississippi's subsurface, and these reflect the tensional forces that developed in the crust as Pannotia was torn apart. A graben is a linear valley that drops down between two normal faults. Grabens are topographically low areas that often feature lakes and rapidly fill with sediment. Numerous grabens developed along the Alabama-Oklahoma Transform Fault, a major northwest-southeast-trending fault that dissected Mississippi. The Mississippi Valley Graben and those associated with the breakup of Pannotia forever weaken the foundation of the North American continent.

Four continents developed from the rifted Pannotia; the two related to our discussion are Laurentia (today's North America, Scotland, and Greenland) and Gondwana (today's South America, Africa, southern Eurasia, Australia, and Antarctica). As these two continents drifted apart the proto-Atlantic Ocean, or Iapetus, developed. When the rifting ended, the southern coast of Laurentia became tectonically quiet, what is called a *passive continental margin*. Central Mississippi and Alabama, part of this coastline, formed a large promontory that extended into Iapetus.

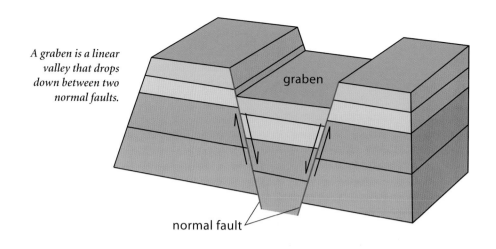

A graben is a linear valley that drops down between two normal faults.

graben

normal fault

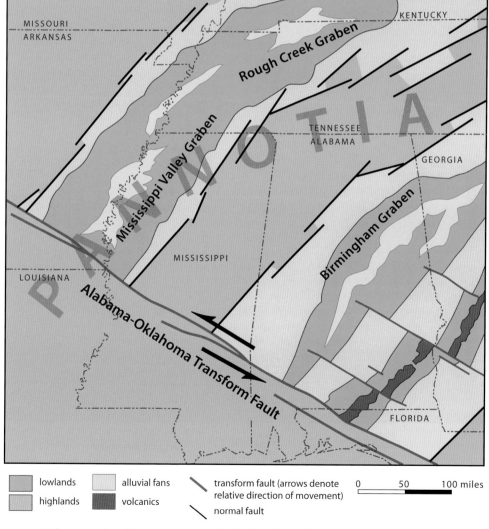

MISSOURI
ARKANSAS

KENTUCKY

Rough Creek Graben

PANNOTIA

Mississippi Valley Graben

TENNESSEE
ALABAMA

GEORGIA

Birmingham Graben

MISSISSIPPI

LOUISIANA

Alabama-Oklahoma Transform Fault

FLORIDA

| | lowlands | | alluvial fans | transform fault (arrows denote relative direction of movement) | 0 | 50 | 100 miles |
| | highlands | | volcanics | normal fault | | | |

Paleogeography of the southern states during late Precambrian time, as Pannotia rifted apart. (Modified in part after Thomas 1991 and Salvador 1991.)

Throughout most of Paleozoic time thousands of feet of carbonate rocks composed of limestone and dolomite formed on Laurentia's continental shelf, which included the northern half of Mississippi. Shallow continental shelves in tropical areas that receive little clastic input (sand and clay sediment) are prime locations for the deposition of carbonate rocks and for the creation of carbonate banks, which are important environments for warm, shallow-water organisms, such as corals, bivalves, mollusks, and other marine invertebrates.

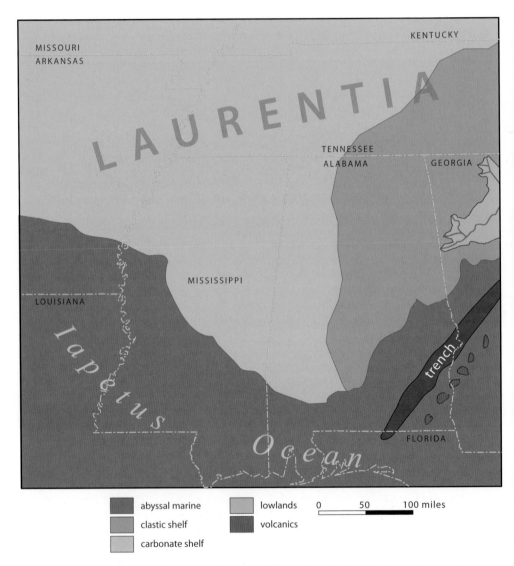

Paleogeography of the southern states during Middle to Late Ordovician time, when northern Mississippi was part of Laurentia's continental shelf. The oceanic trench off the east coast marks the position of the subduction zone responsible for volcanic activity during this time. (Modified in part after Thomas 1991 and Salvador 1991.)

Mississippi's continental-shelf environment would have been similar to that of the Yucatán Peninsula today, which is marked by clear, warm water. Calcite and dolomite, the primary minerals in carbonate rocks, are not stable in cold water. So as water depth increases and temperatures drop, those minerals dissolve. In deep water off the continental shelf, or the abyssal environment, organic-rich clay sourced from other continents was the dominant sediment deposited.

On a global scale, ocean basins open and close on a roughly 400-million-year cycle. The Iapetus Ocean reached its climax during Devonian time and then slowly disappeared, as if a giant zipper first closed in Newfoundland and gradually wrapped its way southwestward around the southern margin of Laurentia, what we call Texas today. As the Iapetus Ocean closed, oceanic crust was consumed along a subduction zone that paralleled the eastern and southern margins of the continent. Volcanoes formed at the surface above the subduction zone as magma was generated from the melting oceanic crust. Large sheets of Cambrian- and Ordovician-age limestones were thrust to the northwest, the north, and finally the northeast due to the ongoing compression related to the closing of the ocean, creating the ancestral Appalachian and Ouachita Mountains. Mississippi was ground zero for this mountain building. Rocks southeast of Mississippi were thrust northwest as Laurentia collided with what is today Africa, and rocks to the south were thrust north-northeast as Laurentia collided with what is today South America. This compressional upheaval lasted for approximately 250 million years and formed mountains as impressive as the Himalayas.

As Iapetus closed, the sedimentation in Mississippi changed from limestone and dolomite (carbonates) to primarily sand and clay (clastics) eroded from the newly created highlands. The oldest rock exposed in Mississippi is the Ross Formation, which was deposited on a carbonate shelf during the Devonian Period (419 to 359 million years ago) in the extreme northeastern part of the state. (You can see the Ross limestone along the shore of Pickwick Lake, but it is not featured in a road log.) Lying just above the Ross Formation are the Mississippian-age Fort Payne Chert and Tuscumbia Limestone. These rocks represent the last of the significant carbonate deposition that occurred in Mississippi before Iapetus closed. Both formations are exposed in Tishomingo County, but the best exposures are in quarries and along the shores of Pickwick Lake and the Tennessee River and its tributaries.

Rivers dumped countless tons of sediment—generated from the rapidly eroding ancestral Appalachian Mountains—in the Black Warrior Basin. This triangular-shaped depression (now buried), situated south of Tennessee and centered across the Mississippi-Alabama border, was part of the Iapetus Ocean. The weight of the sediment caused the crust to depress, allowing for a huge volume to be deposited in the basin. The total combined thickness of Mississippian and Pennsylvanian sediments deposited in the basin exceeds 15,000 feet.

During Mississippian time, most of the sediment was sourced from the north and northeast, from the Appalachians. In northeastern Mississippi today, the Hartselle Sandstone and Pride Mountain Formation, both of Mississippian age, are exposures of the Black Warrior Basin sediments. The Pride Mountain

Formation's interbedded shales, sands, and limestones were deposited in and around the margins of the shallow Iapetus as it closed. The Hartselle Sandstone is interpreted to represent a barrier island complex, such as that which has developed off the coast of Mississippi today.

During Pennsylvanian time, 323 to 299 million years ago, the Black Warrior Basin of northern Mississippi was still covered by a shallow sea, but western and southern Mississippi were mountainous. As South America converged on what would become Mississippi, northwest-southeast-trending thrust faults

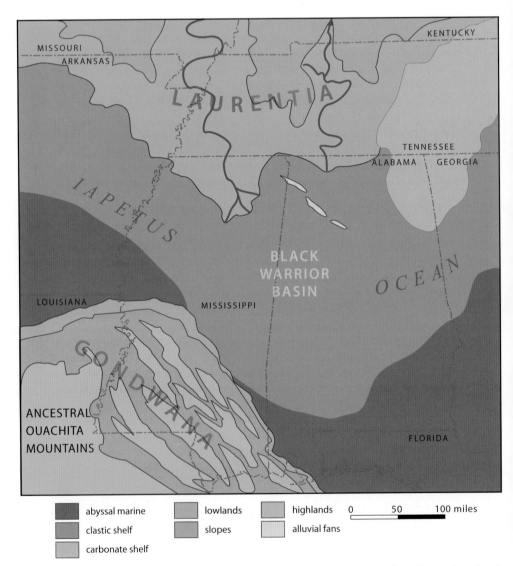

abyssal marine		lowlands		highlands		
clastic shelf		slopes		alluvial fans		
carbonate shelf						

0   50   100 miles

Paleogeography of the southern states during Late Mississippian time. Much of the sediment deposited in the Black Warrior Basin was sourced from the eroding ancestral Appalachians to the north. (Modified after Allmon, Picconi, Greb, and Smith 2011 and Salvador 1991.)

created what is known as the Frontal Ouachita Mountains in the central part of the state. As mountains (the ancestral Ouachitas in western Mississippi and Arkansas, the Frontal Ouachitas in central Mississippi, and the ancestral Appalachians in Alabama) rose all around the perimeter of the Black Warrior Basin, sediments began to be sourced from them. The depositional environments in northern Mississippi were transitioning from those dominated by marine processes to those dominated by terrestrial.

The development of these coastal environments was timed perfectly with the evolution of abundant land plants. The river systems flowing from the mountains created large deltas and wetlands, both ecosystems that support the

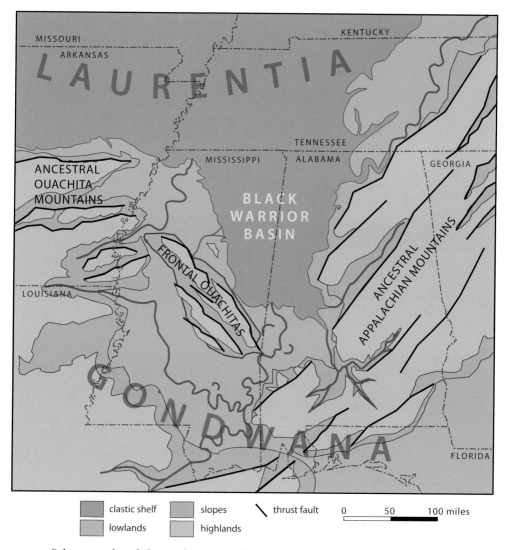

Paleogeography of the southern states during Pennsylvanian time. (Modified after Picconi, Greb, and Smith 2011 and Salvador 1991.)

growth of abundant vegetation. With burial, the wetland vegetation was preserved and then compacted to eventually form the Black Warrior Basin's coal seams. Though mined elsewhere in the United States, the Pennsylvanian-age coals in Mississippi are not mined due to their depth. By the end of the Permian period 252 million years ago, Iapetus had closed completely and the continents had coalesced as Pangaea, Earth's most recent supercontinent.

Because the Earth never rests, Pangaea did not stick together for long. Approximately 200 million years ago, during Triassic time, Pangaea began to break up, and 25 million years later the Gulf of Mexico was born. Mississippi was rifted along a zone that paralleled the ancient Alabama-Mississippi Transform Fault. The difference between the opening of Iapetus and the opening of the Gulf of Mexico is that this time the tensional forces were oriented northeast-southwest rather than northwest-southeast.

Besides normal and transform faults, which are part and parcel of a continent being torn in two, geologists look for other clues when identifying a continental rift. Alluvial fans form where sediments eroded from mountainous areas travel only a short distance and are deposited at the base of the mountains. (The Eagle Mills Formation in Mississippi was deposited as alluvial fans during the rifting of Pangaea. However, it is not exposed at the surface.) The grabens that develop at rifts are steep sided and excellent sites for alluvial fans to accumulate. Lake sediments are also a clue for former rifting; as noted earlier, lakes often form in the valleys of grabens. As Earth's crust is stretched in rift zones it grows thinner; thus, volcanoes form in rift valleys as magma seeks a way to the surface along paths of least resistance. So volcanic rocks are also clues to former rifting.

What happened in Mississippi 200 million years ago is similar to what is occurring in eastern Africa today. There the African Plate is tearing in two, forming the East African Rift, a series of rift valleys that are home to numerous lakes, such as Victoria, Turkana, and Tanganyika, and also the volcanoes Mount Kilimanjaro and Mount Kenya. The Triassic rifting was not very clean, meaning the continents didn't break apart exactly along preexisting seams. Parts of Gondwana were left attached to the Mississippi and Alabama portions of Laurentia. One of these, locally known as the Wiggins Uplift, remained a component of the south Mississippi landscape. The Wiggins Uplift underlies the six southernmost counties and is composed of metamorphosed granite.

During Triassic, Jurassic, and Cretaceous time (252 to 66 million years ago), the Appalachian and Ouachita Mountains eroded to a relatively smooth surface. Mississippi formed part of the coast along a tectonically quiet, passive continental margin. North and south of the Wiggins Uplift, the crust continued to rift apart and subside. Whereas the rest of southern Mississippi eventually succumbed to the waters of the nascent Gulf of Mexico, this granite island remained above sea level throughout most of Jurassic time.

The area north-northwest of the uplift became the Mississippi Interior Salt Basin. The Jurassic-age Werner Formation salt and Louann Salt, deposited between 170 and 161 million years ago, represent the first widespread sedimentation in the newly created basin. As previously mentioned, subsidence plays a

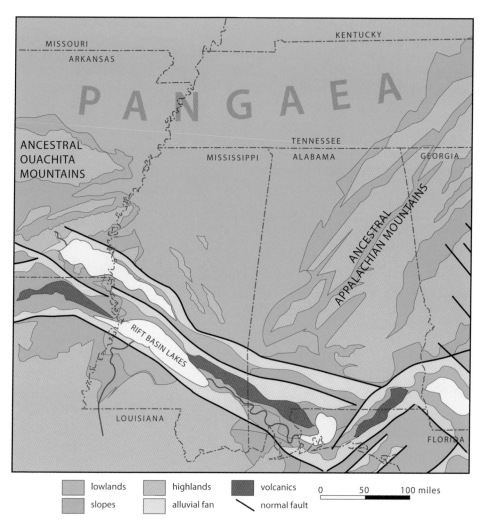

Paleogeography of the southern states during Triassic time. The Appalachian and Ouachita Mountains continue to erode, but the big story is the breakup of Pangaea. The northwest-southeast rift was home to lakes, alluvial fan sedimentation, and volcanoes. (Modified after Salvador 1991.)

major role in sedimentary geology. Unnamed rivers carried sediment from the midcontinent area of North America and formed deltas in west-central Mississippi. The weight of sediment delivered to the coast, and that of the carbonates that formed on the continental shelf, caused the crust to subside, which allowed the coast to accommodate more sediment. In the deepest parts of the Mississippi Interior Salt Basin, the base of the salt is more than 30,000 feet below the surface. The accumulation of this much sediment would not have been possible had the crust not sagged under the sediment's weight.

Following the deposition of the Louann Salt, the Norphlet Formation sand-stone, Smackover Limestone, and Haynesville Formation (sand, shale, and evaporite) were deposited. *Evaporites* are minerals that precipitate from water as it evaporates; they collect to form sedimentary rocks, including halite (salt) and gypsum. These formations, deposited in marine and terrestrial environments, ranging from abyssal to coastal dunes, are not exposed at the surface in Mississippi.

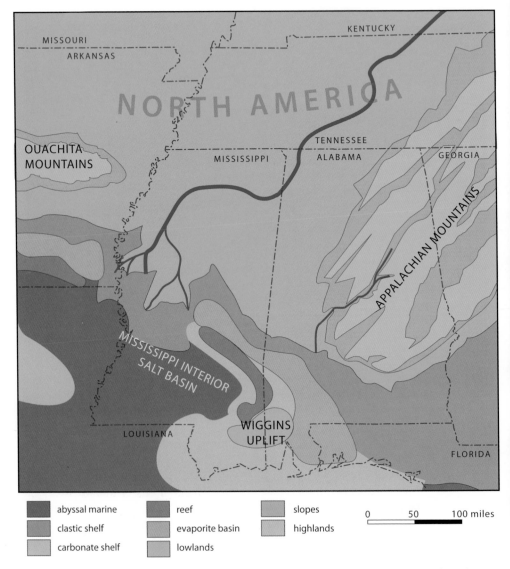

Paleogeography of the southern states during Late Jurassic time. Salt was precipitating from the shallow sea and filling the Mississippi Interior Salt Basin and the developing Gulf of Mexico. (Modified after Salvador 1991.)

As sea level rose and the amount of sand and clay eroded from the mountains and delivered to the newly created basin increased, the crust continued to sag under the weight. These sediments would become the rocks of the Jurassic-age Cotton Valley Group and numerous Cretaceous-age units, starting with the Hosston Formation and ending with the Washita-Fredericksburg Formation; none are exposed at the surface. Following the deposition of the Washita-Fredericksburg sediment, sea level dropped and a period of erosion occurred in

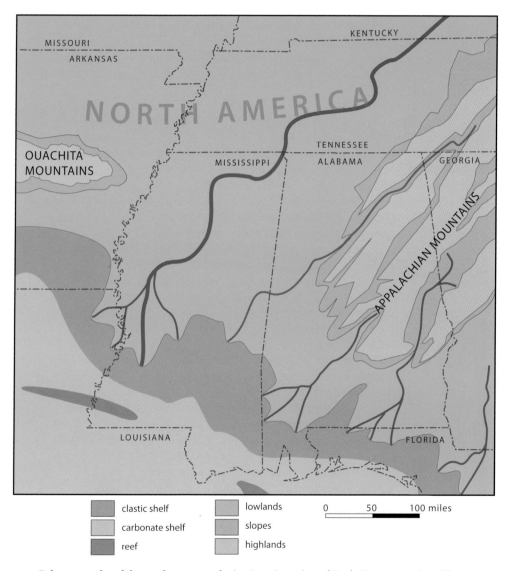

Paleogeography of the southern states during Late Jurassic and Early Cretaceous time. Rivers continue to deliver sediment sourced from the Appalachians to the nascent Gulf of Mexico, and carbonates precipitate from the sea in southern Mississippi. (Modified after Salvador 1991.)

most of Mississippi. When sea level rose again, the sand and shale of the Coker and Gordo Formations, both of the Tuscaloosa Group, covered the state with a veneer of sediment. In the northern third of the state the Tuscaloosa Group overlies Paleozoic-age rocks that were exposed during erosion.

Following another period of sea level drop and erosion across the state, sea level rose again, and the sands of the Eutaw Group were deposited in coastal and delta environments along the coast in northern Mississippi. In southern Mississippi the Eutaw Group transitioned into chalk. As sea level continued to rise, a broad, shallow sea formed. Chalk deposition spread northward, resulting in the formation of the Selma Group. The chalks of the Selma Group are the Mooreville, Demopolis, and Prairie Bluff Formations, from the bottom up. The Demopolis Formation attained the greatest extent and can be found in southern Tennessee. As chalk was deposited on the continental shelf, sand and clay of the Coffee Sand and Ripley Formation were deposited in northern Mississippi along the coastline.

The Cretaceous sea was not a completely tranquil place; there was some anomalous volcanic activity in the center of the basin around 77 million years ago. Geologists have proposed several theories to explain the igneous activity. One calls for the North American Plate passing over a feature known as the Bermuda hot spot (see "The Jackson Dome" on page 44). Whatever the cause, there were several volcanoes in the Cretaceous sea.

In Mississippi there were the Midnight and Jackson volcanoes. The Jackson volcano, which formed the Jackson Dome, pierced the entire thickness of sediment in the Mississippi Interior Salt Basin, doming up some of the rock formations with it. The volcano weathered and eroded to form an island roughly 15 miles in diameter. Rocks ranging from the Jurassic-age Cotton Valley Group, in the center of the island, up through the Cretaceous-age Eutaw Group, on its flanks, were exposed. The Selma Chalk deposited on top of the eroded volcanic rock is called the Jackson Gas Rock. It is the reservoir rock for the Jackson Gas Field, which was an important source of natural gas for the capital city of Jackson during the Great Depression.

The sediments deposited during Paleogene time (66 to 23 million years ago) and Neogene time (23 to 2.6 million years ago) make up the majority of the rocks exposed at the surface of Mississippi. All of these formations were deposited in the Mississippi Embayment as the ocean retreated from the North American continent. The embayment was a linear depression, or basin, that extended from the Gulf of Mexico to southern Illinois, near Cairo (see "Mississippi Embayment" on page 36 for more information). Mississippi was at or near the shoreline throughout these geologic periods of time.

Geologists refer to sea level changes in relative terms, that is, whether the level of the water is rising or falling relative to the land surface. On a global scale, with the exception of a sea level high between 56 and 48 million years ago, during the Eocene Epoch, sea level has dropped since the beginning of Paleogene time. However, within this long-term trend there were numerous complex and intermittent sea level fluctuations over the short term. When there was a relative rise in sea level (transgression), marine environments shifted landward, resulting in the deposition of marine clay, marl, and limestone; when sea level fell (regression), terrestrial environments shifted toward the ocean, resulting in the advance of large delta systems. The Paleogene-age formations in Mississippi are characterized by the cyclical deposition of marine and delta deposits, whereas the Neogene-age formations were more terrestrial in nature, dominated by river and delta deposits. Changes in sea level are caused by many factors, including a rise or fall in land surface, changes in the volume of the

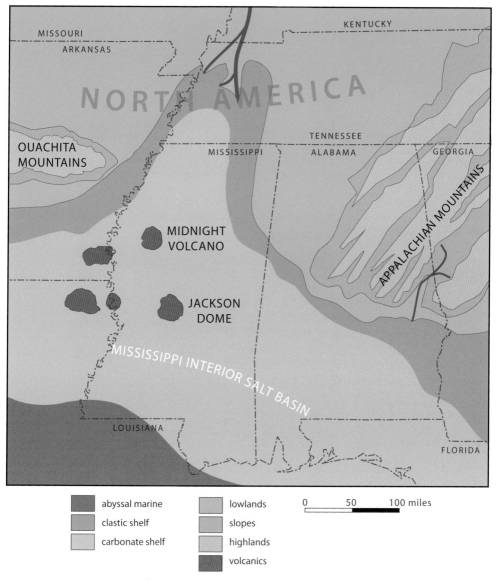

Paleogeography of the southern states during Cretaceous time. Several volcanoes developed in the Mississippi Interior Salt Basin. (Modified after Salvador 1991.)

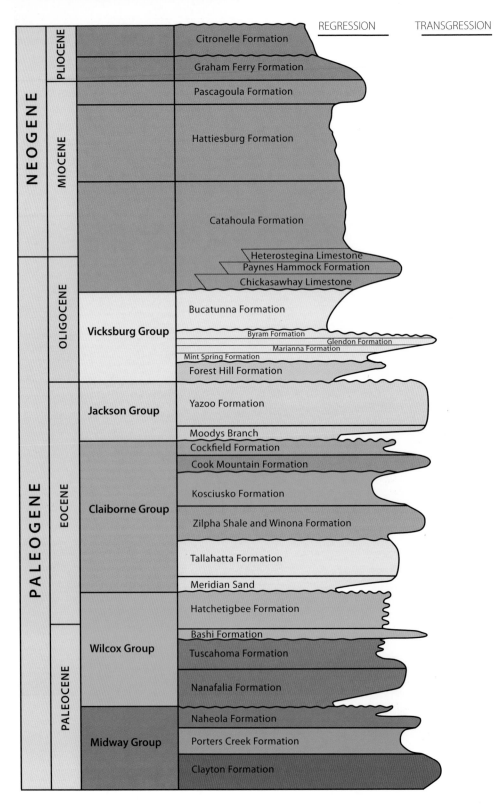

Paleogene and Neogene sea level curve for the Mississippi Embayment, and the resulting formations. (Modified after Dockery 1986.)

N E O G E N E	Citronelle Formation Graham Ferry Fm.	The **Graham Ferry Formation** is composed of sand and clay deposited on a delta as sea level gradually dropped. Clay, sand, and gravel of the **Citronelle Formation**, the last sediments deposited in the Neogene, dominate southern Mississippi. The sediments were deposited in nearshore environments to those of broad, braided river channels as ancient rivers scoured north and central Mississippi and the Gulf of Mexico receded to approximately where it is today.
	Pascagoula Formation Hattiesburg Formation Catahoula Formation Chickasawhay Ls. Paynes Hammock Fm. Heterostegina Ls.	The sand and clay of the **Catahoula, Hattiesburg,** and **Pascagoula Formations** were deposited in coastal and delta environments along Mississippi's southern coast. A minor transgression deposited marine sand of the **Pascagoula Formation**. The **Chickasawhay, Paynes Hammock,** and **Heterostegina limestones** were deposited offshore in eastern Mississippi.
	VICKSBURG GROUP Bucatunna Formation Byram Formation Glendon Limestone Marianna Limestone Mint Spring Formation Forest Hill Formation	A significant drop in sea level resulted in the coastal plain deposits of the **Forest Hill Formation** on top of the weathered **Yazoo Clay**. The shallow-marine limestone, marl, and sand of the **Mint Spring Formation** and **Marianna** and **Glendon Limestones** record a complete marine transgression followed by a regressive sequence composed of the **Byram Formation** sand and marl and **Bucatunna Formation** clay. The **Marianna** and **Glendon Limestones** represent the last major transgressions in the Gulf of Mexico.
P A L E O G E N E	JACKSON GROUP Yazoo Clay Moodys Branch Fm.	The transgressive sea that deposited the **Moodys Branch** marl on top of the eroded surface of the **Cockfield Formation** eventually culminated with the deposition of the **Yazoo Clay** and also some marl.
	CLAIBORNE GROUP Cockfield Formation Cook Mountain Fm. Kosciusko Formation Zilpha Shale Winona Formation Tallahatta Formation	The next major surge of terrestrial deposition came with the Claiborne Group. The **Tallahatta, Winona, Kosciusko,** and **Cockfield Formations** were all deposited in shallow-marine to nearshore and delta environments. The mineral composition of **Claiborne** sands suggests that the dominant sediment source had shifted away from the Appalachian Mountains to a river system flowing from a midcontinental source. The **Zilpha Shale** and **Winona** and **Cook Mountain Formations** record major transgressions over the continental shelf and deeper-water environments. The transgression that led to the deposition of **Cook Mountain** clay and silt was the most extensive in the Claiborne Group.
	WILCOX GROUP Hatchetigbee Fm. Bashi Formation Tuscahoma Formation Nanafalia Formation	The **Wilcox Group** was deposited on a coastal plain and delta that developed in southern Mississippi and records many small fluctuations in sea level. The most notable transgressive event is associated with the very fossiliferous **Bashi Formation**, which was deposited as sea level rose and the marine environment swept landward. Studies of heavy minerals in the **Wilcox Group** suggest the eroding Appalachian Mountains were the source of its sediments, delivered to the embayment by an ancestral river.
	MIDWAY GROUP Naheola Formation Porters Creek Fm. Clayton Formation	The first transgressive episode of the Paleogene began with the **Clayton** and **Porters Creek Formations**. The **Clayton Formation** is composed of sandy limestone, chalk, and marl that were deposited on a shallow shelf throughout much of Mississippi. The **Porters Creek** clay was deposited in a deeper-water shelf environment. As sea level receded, the sands of the **Naheola Formation** were deposited by coastal streams moving over the resulting floodplain.

Geologists have identified numerous transgressive and regressive cycles in Mississippi's Paleogene and Neogene Formations, the formations most often seen at the surface in Mississippi.

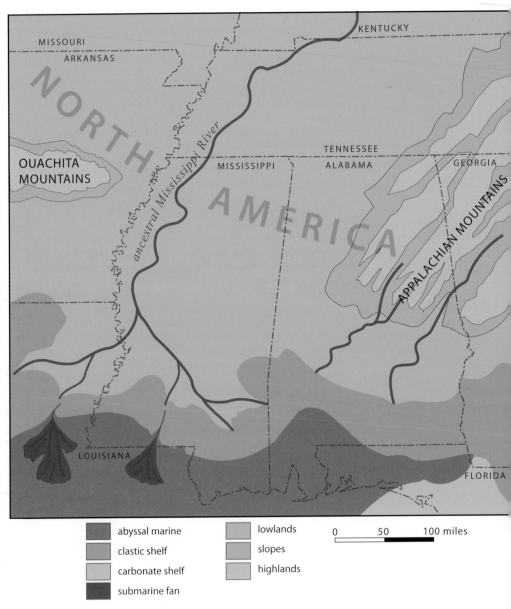

MISSOURI

ARKANSAS

KENTUCKY

NORTH

OUACHITA
MOUNTAINS

ancestral Mississippi River

TENNESSEE

MISSISSIPPI

ALABAMA

GEORGIA

AMERICA

APPALACHIAN MOUNTAINS

LOUISIANA

FLORIDA

abyssal marine

lowlands

clastic shelf

slopes

carbonate shelf

highlands

0   50   100 miles

submarine fan

Paleogeography of the southern states during Paleocene time, a period when seas were rising and falling across Mississippi. During periods of lower sea level the ocean receded from the Mississippi Embayment. Deltas at the shoreline and submarine fans just seaward of the continental shelf were significant features that formed as sediment was delivered to the coast. (Modified after Salvador 1991 and Saucier 1994.)

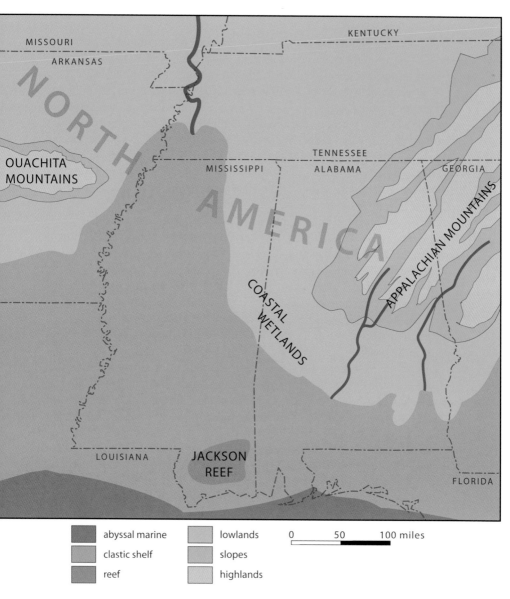

MISSOURI

ARKANSAS

KENTUCKY

NORTH

OUACHITA
MOUNTAINS

TENNESSEE

MISSISSIPPI

ALABAMA

GEORGIA

AMERICA

APPALACHIAN MOUNTAINS

COASTAL
WETLANDS

LOUISIANA

JACKSON
REEF

FLORIDA

	abyssal marine		lowlands
	clastic shelf		slopes
	reef		highlands

0   50   100 miles

Paleogeography of the southern states during the Eocene transgression that inundated the Mississippi Embayment one more time and deposited the Yazoo Clay or its equivalents throughout Mississippi. (Modified after Salvador 1991 and Saucier 1994.)

ocean basin due to the swelling of a mid-ocean ridge, or changes in the volume of water due to temperature (warmer seawater has more volume).

The Quaternary Period ranges from 2.6 million years ago to the present. Within that timeframe was the Pleistocene Epoch (2.6 million years ago to 11,700 years ago), not the last episode in Mississippi's geologic history but an important one. The Pleistocene is known for glacial growth (advances) and glacial melting (retreats). Although ice never came close to Mississippi, it left its mark on the state.

Prior to the Pleistocene the ancient Teays River system drained runoff from the states of Kentucky, Virginia, West Virginia, Ohio, and Indiana. The river flowed to the west through Illinois and then south to the Gulf of Mexico. Glacial advances destroyed much of the river's channel, causing water to flow in different directions and new rivers to form altogether. Research indicates that as early as 85,000 years ago the ancestral Ohio River formed. It flowed through parts of the old Teays channel but went southward rather than westward, as the Teays had. The Ohio competed with the Mississippi River for floodplain in the Mississippi Embayment. So, many of the oxbow lakes, channels, and meander scars we see in the Mississippi Delta could be from either the ancestral Mississippi or the ancestral Ohio.

Before the ancestral Ohio and Mississippi Rivers became entrenched in their modern alluvial valleys, they left vast gravel deposits across the Delta region of western Mississippi. Today this gravel is exposed, in places, more than 100 feet above the floodplain in the modern Mississippi's bluffs.

During the maximum extent of the last glacial advance, approximately 22,000 years ago, sea level in the Gulf of Mexico dropped 410 feet because so much water was frozen as ice across the globe. This drop in sea level increased the gradient of the ancestral rivers, which caused them to erode into the Neogene- and Paleogene-age sediments of the Mississippi Embayment, creating a valley. As the ice retreated and sea level rose, the ancestral rivers filled some of this scoured-out valley with gravel and sand that the melting glaciers had entrained and piled up, to the north. This thick accumulation of sand and gravel formed the great Mississippi River Valley Alluvial Aquifer (discussed in more detail in "The Delta" on page 108).

As glaciers advanced and retreated throughout the Pleistocene Epoch, they ground and pulverized the rocks beneath them, creating silt-sized particles called *glacial flour*. As the prevailing westerly winds of the Pleistocene swept across the central United States, glacial flour was picked up, carried eastward, and deposited as loess along the newly exposed bluffs along the river. The modern configuration of the Ohio and Mississippi Rivers we are familiar with was established approximately 10,000 years ago. The meandering Mississippi River scoured the landscape in western Mississippi from Memphis to Vicksburg and left behind the fertile soils of the Delta. The bluffs today offer spectacular views out over this region.

The subtle nature of Mississippi's geology at the surface in no way reflects the dynamic events that occurred here in the past. The record of coastal sedimentation and Mississippi floodplain deposits seen on the geologic map of the state portrays a very different geologic history than that seen in cross section.

When you slice an orange in half you are able to see the interior structure of the fruit in what is called *cross section*. Geologists can create cross sections of the Earth by interpreting and correlating records created during seismic

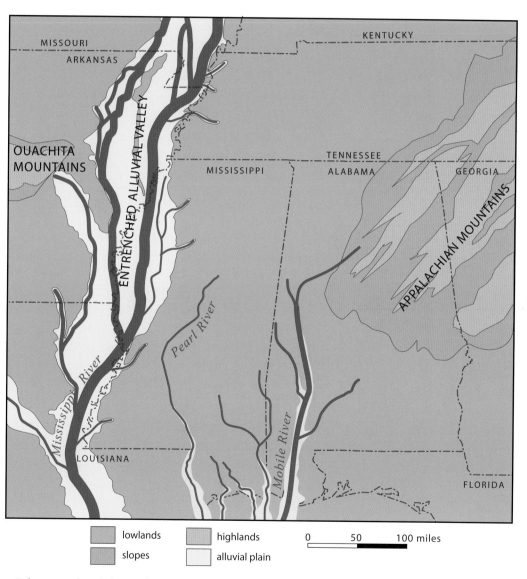

Paleogeography of the southern states during Pleistocene time. The Mississippi River establishes its course, depositing the fine sediment of the fertile Mississippi River Alluvial Plain (yellow), or the Delta. Prevailing westerly winds deposit loess along the eastern bluffs of the plain. (Modified after Saucier 1994.)

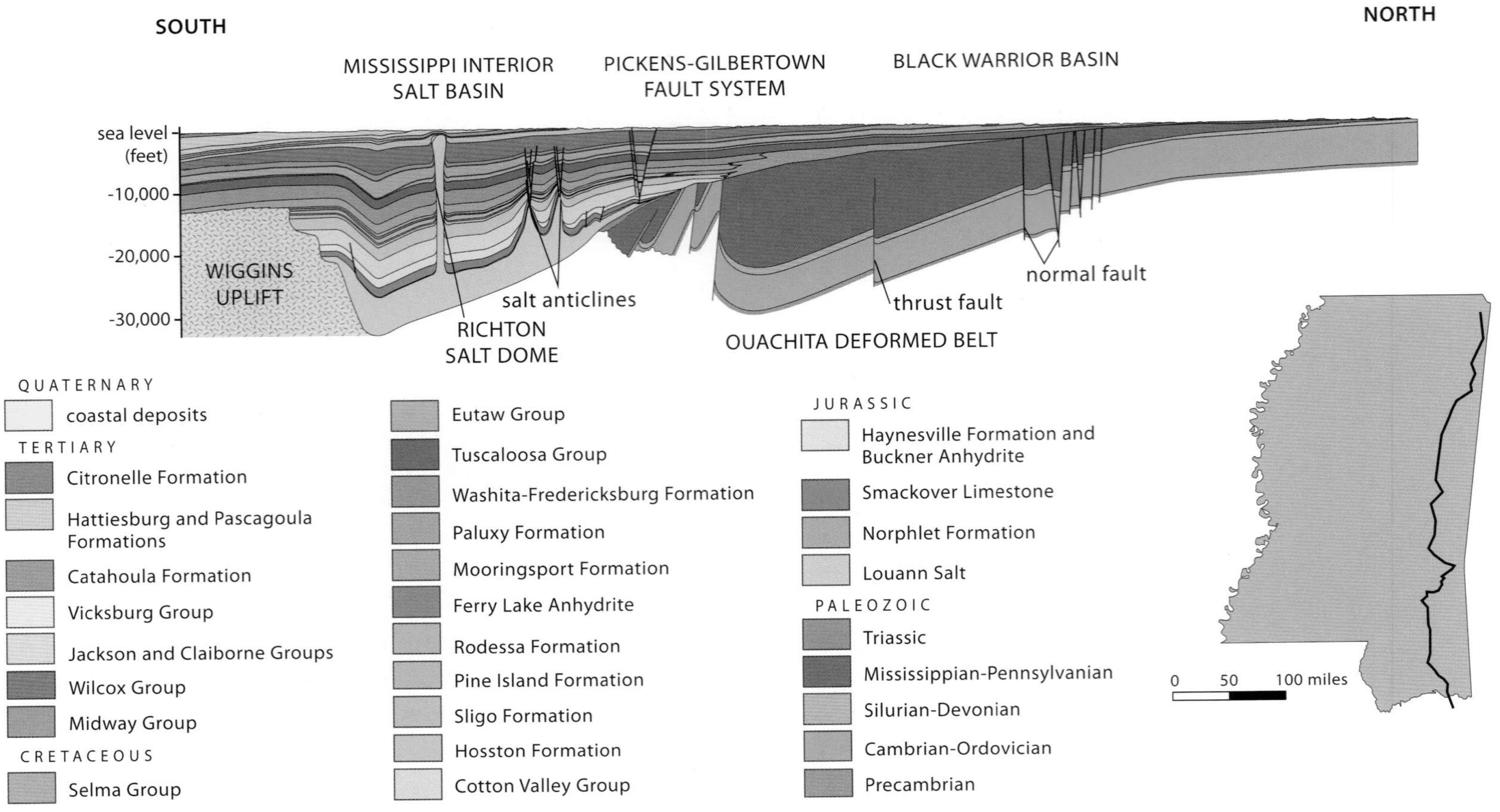

SOUTH — NORTH

sea level (feet)
-10,000
-20,000
-30,000

MISSISSIPPI INTERIOR SALT BASIN

PICKENS-GILBERTOWN FAULT SYSTEM

BLACK WARRIOR BASIN

salt anticlines

RICHTON SALT DOME

WIGGINS UPLIFT

OUACHITA DEFORMED BELT

thrust fault

normal fault

0 50 100 miles

QUATERNARY
coastal deposits

TERTIARY
Citronelle Formation
Hattiesburg and Pascagoula Formations
Catahoula Formation
Vicksburg Group
Jackson and Claiborne Groups
Wilcox Group
Midway Group

CRETACEOUS
Selma Group
Eutaw Group
Tuscaloosa Group
Washita-Fredericksburg Formation
Paluxy Formation
Mooringsport Formation
Ferry Lake Anhydrite
Rodessa Formation
Pine Island Formation
Sligo Formation
Hosston Formation
Cotton Valley Group

JURASSIC
Haynesville Formation and Buckner Anhydrite
Smackover Limestone
Norphlet Formation
Louann Salt

PALEOZOIC
Triassic
Mississippian-Pennsylvanian
Silurian-Devonian
Cambrian-Ordovician
Precambrian

North-south cross section of Mississippi. (Modified after Williams 1969.)

exploration and the drilling of oil wells. The north-south cross section of Mississippi reveals how complex Mississippi's geologic history has been. Be aware that this cross section is greatly exaggerated; if it were drawn to true scale, none of the features would be visible because only a very thin line would represent the 30,000 feet of sediments.

Starting in the north there is a gradual dip of Paleozoic-age rocks southward into the Black Warrior Basin, in which more than 15,000 feet of Mississippian- and Pennsylvanian-age sediment was deposited. The collision with South America (Gondwana) generated thrust faults, which constitute the Ouachita Deformed Belt. The collision forced those Paleozoic rocks northeast along thrust faults, completing the construction of the ancestral Ouachita Mountains during Pennsylvanian time and signaling the end of mountain building in Mississippi. That mountain range, which developed between Jackson and Memphis, was eroded and transported to the ocean during the next 200 million years. The depressed area south of the Black Warrior Basin is the Mississippi Interior Salt Basin. Subsidence allowed more than 30,000 feet of sediment to accumulate after the Gulf of Mexico began to open 175 million years ago. The Richton Salt Dome and two additional faulted salt anticlines (see "Salt Domes" on page 180 for more information) formed in this basin. It is apparent in several places where the salt movement folded rocks upward. The Pickens-Gilbertown Fault System is a zone of normal faulting that formed as the salt basin subsided throughout the Mesozoic Era. The Wiggins Uplift, a chunk of subsurface granite, is a portion of South American crust that was left behind as the modern continents separated during the rifting of Pangaea. It was not until the Late Jurassic that sediment was deposited on top of the feature. Though the modern Gulf of Mexico, the most recent basin to form in Mississippi, is not visible, the slight dip of the Eocene-age Jackson and Claiborne Groups hint at the basin to the south.

PHYSIOGRAPHIC PROVINCES

There are nine physiographic provinces in the state of Mississippi. A *physiographic province* is a geographic area with similar bedrock, geologic structure, and topographic expression, all of which are directly related to geology. Rock type and structure (whether the rocks are faulted, folded, or horizontal) determine how weathering and erosion affect bedrock. The gentle dip of the rocks in the state to the west and southwest, into the Mississippi Embayment, does influence where physiographic boundaries are located, but rock type and composition are more significant. Bedrock controls most changes in topography, soil color, and vegetation.

The oldest rocks in Mississippi are exposed where the Tennessee and Tombigbee Rivers have eroded down through Mesozoic-age rocks. This province, the Paleozoic Bottoms, is only exposed in Tishomingo County, and its physiography is expressed over the Devonian-age Ross Formation and Chattanooga Shale; the Mississippian-age Fort Payne Chert, Tuscumbia Limestone, and Hartselle Sandstone; and the Pride Mountain Formation. The Chattanooga

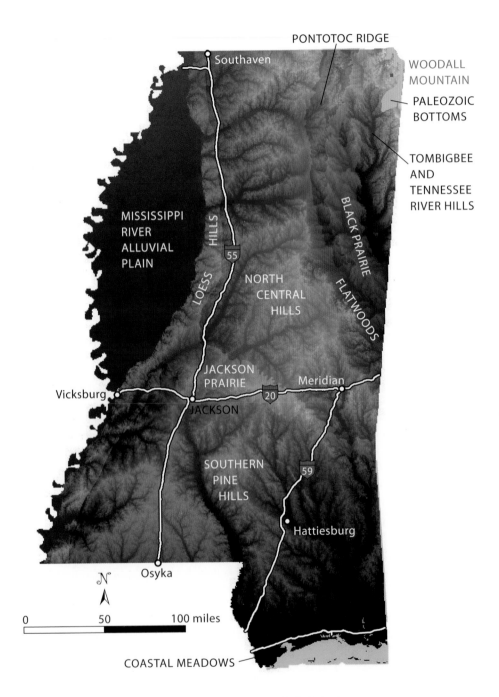

PONTOTOC RIDGE

WOODALL
MOUNTAIN

PALEOZOIC
BOTTOMS

TOMBIGBEE
AND
TENNESSEE
RIVER HILLS

Southaven

MISSISSIPPI
RIVER
ALLUVIAL
PLAIN

LOESS HILLS

55

NORTH
CENTRAL
HILLS

BLACK PRAIRIE

FLATWOODS

JACKSON
PRAIRIE

Meridian

Vicksburg

20

JACKSON

SOUTHERN
PINE
HILLS

59

Hattiesburg

N

Osyka

0   50   100 miles

COASTAL MEADOWS

Physiographic provinces of Mississippi.

Shale and the Pride Mountain Formation are composed of shales that erode relatively easily, so they tend to form low-angle slopes. The Ross, Fort Payne, and Tuscumbia limestones and the Hartselle Sandstone are better cemented and thus resistant to erosion, so they tend to form ridges.

The sands of the Cretaceous-age Tuscaloosa and Eutaw Groups and the Coffee Sand underlie the Tombigbee and Tennessee River Hills physiographic province. This province features the highest elevation in Mississippi: 806-foot-tall Woodall Mountain, which is capped by Coffee Sand. Due to their porous nature, precipitation rapidly infiltrates these formations, which limits storm-water runoff and its erosive effects.

The Black Prairie was originally named for its dark soils and clays that formed as chalks of the Selma Group weathered. The change in physiography at the boundary of the Black Prairie and Tombigbee and Tennessee River Hills developed where the Mooreville Formation chalk of the Selma Group transitions into the Coffee Sand. The chalk weathers chemically and is easily dissolved, creating more rolling hills, whereas the Coffee Sand tends to erode via running water; it forms more rugged hills and valleys. The Pontotoc Ridge physiographic province features the second-highest hills in Mississippi; sands of the McNairy and Chiwapa Members of the Ripley and the Owl Creek Formations underlie this region. The southern end of Pontotoc Ridge is located between Starkville and Macon, where the Ripley Formation transitions into chalk.

The Flatwoods and Jackson Prairie physiographic provinces are typically low-relief areas that have developed on clay bedrock that erodes relatively easily. Clay of the Paleocene-age Porters Creek Formation of the Midway Group outcrops in a narrow band between the Black Prairie and North Central Hills, forming the Flatwoods province. The Jackson Prairie is defined primarily by gently rolling hills that have developed on the Eocene-age Yazoo Clay of the Jackson Group.

The Eocene-age Claiborne Group and Paleocene-Eocene-age Wilcox Group underlie the North Central Hills physiographic province, and the Southern Pine Hills have developed on the Oligocene-age Vicksburg Group; the Oligocene- to Miocene-age Catahoula, Pascagoula, and Hattiesburg Formations; and the Pliocene-age Citronelle Formation. There are thin limestones in the Vicksburg Group, but these provinces have developed predominantly in sands, the resistant nature of which has led to the rolling aspect of their hills. Citronelle gravels that escaped erosion are found on top of some of the hills in the Southern Hills physiographic province and are often the highest relief.

In western Mississippi the remaining physiographic provinces are the Mississippi River Alluvial Plain and Loess Hills. The latter province is composed of windblown glacial flour that stands 100 to 200 feet above the eastern edge of the Mississippi River Alluvial Plain. The bluffs mark the eastward limit of the Mississippi River's erosional and flood influence, and they are home to numerous steep-sided valleys in which streams have eroded down through the loess. Eastward of the bluffs, the loess-rich soils are good for agriculture due to their porous nature and properties afforded by clay minerals. The Mississippi River

Alluvial Plain is underlain by gravel, sand, silt, and clay deposited by both the ancestral and modern Mississippi River in channels and point bars, and it has the least relief of any province in the state.

Coastal Mississippi, at elevations lower than 60 feet, comprises the Coastal Meadows physiographic province. Its deposits were primarily deposited in alluvial, delta, and shoreline environments. Units include the Pleistocene-age Biloxi, Gulfport, and Prairie Formations, and the Holocene-age alluvial and delta deposits of several rivers.

MISSISSIPPI EMBAYMENT

One of the more prominent features observable on the geologic map of the southern states is the Mississippi Embayment. The embayment is a linear depression, or basin, that extended from the Gulf of Mexico to southern Illinois, near Cairo. It is no coincidence that one of the youngest features on the continent, the Mississippi River, meanders its way through the center of the embayment. The weakness in the continental crust flooring the embayment dates back to the failed rift created in Precambrian time, when the North American continent (then part of the larger landmass known as Rodinia) was pulling apart. The rifting subsided, but the series of faults it created, known as the Mississippi Valley Graben, weakened the foundation of North America forever and enabled the crust to subside beneath the weight of overlying sediment. The Mississippi Embayment is like a crack in a plaster wall: you can cover the crack and paint over it, but unless you repair the foundation below (the graben in this case), the weakness will always be there.

Starting approximately 90 million years ago, uplift and magmatic activity, possibly related to a hot spot, further weakened the embayment. Erosion, cooling, and subsidence followed this igneous activity. The weakened zone of the embayment then became the locus of deposition for sediment eroding from ancient highlands (the Ouachita and Appalachian Mountains) as an ocean retreated from the continent via the embayment itself. Most of the rocks seen at the surface in Mississippi were deposited in the embayment as it subsided and then later was exposed by erosion.

The embayment is what geologists call a *plunging syncline*. A *syncline* is a large-scale, convex-down fold in rock strata in which all the strata dip toward the center, or axis, of the structure. In outcrop, the ages of the rock formations get progressively younger as you move toward the center of the fold. The axis of the embayment—the syncline—is oriented in a south-southwest direction, and the depth of any rock formation gets progressively deeper in this direction; that's the "plunge" part of the syncline structure, as in the axis of the syncline *plunges* deeper into the Earth in a south-southwest direction. As a result of the syncline, rock units in north Mississippi, such as those of the Tuscaloosa, Eutaw, Selma, Midway, Wilcox, and Claiborne Groups, have a westerly dip toward the axis of the syncline, which roughly parallels the Mississippi River, whereas the

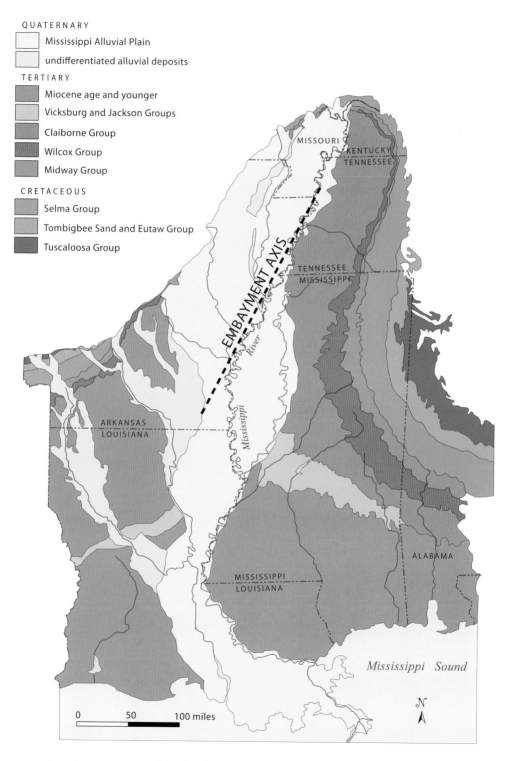

QUATERNARY
- Mississippi Alluvial Plain
- undifferentiated alluvial deposits

TERTIARY
- Miocene age and younger
- Vicksburg and Jackson Groups
- Claiborne Group
- Wilcox Group
- Midway Group

CRETACEOUS
- Selma Group
- Tombigbee Sand and Eutaw Group
- Tuscaloosa Group

MISSOURI

KENTUCKY
TENNESSEE

EMBAYMENT AXIS

TENNESSEE
MISSISSIPPI

ARKANSAS
LOUISIANA

Mississippi River

ALABAMA

MISSISSIPPI
LOUISIANA

Mississippi Sound

0   50   100 miles

N

Geologic map of the Mississippi Embayment. (Modified after Hart, Clark, and Bolyard 2008.)

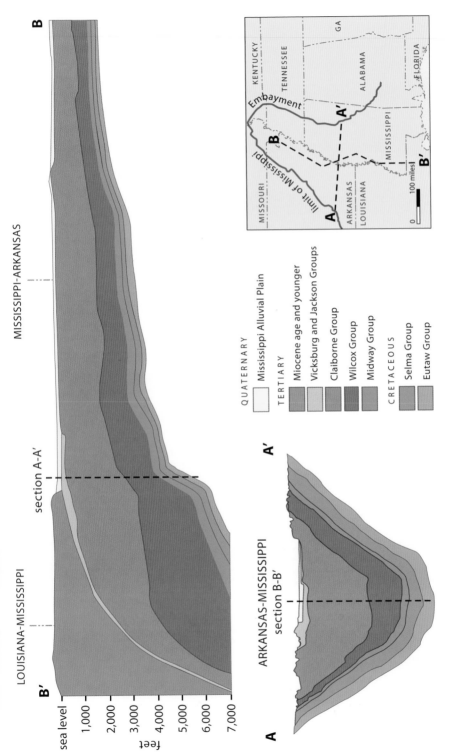

The north-south cross section (top) highlights how the Mississippi Embayment syncline plunges down in a south-southwest direction. The east-west cross section (bottom) shows how the rock units dip toward the axis, or center, of the Mississippi Embayment syncline in central Mississippi. (Modified after Arthur and Taylor 1998.)

rock units south of Jackson, such as those of the Jackson and Vicksburg Groups and younger sediments, have a more southerly dip due to the plunge angle of the syncline itself. Alluvial sediment deposited by the Mississippi River as it meandered across western Mississippi has obscured outcrop patterns in the center of the embayment.

OIL AND GAS

Natural gas was the first type of hydrocarbon discovered in the state in October 1926 with production from the Mississippian-age Carter Sandstone in Amory. But it was the discovery of the Tinsley Oil Field in Yazoo County that really brought significant media attention to Mississippi and set the stage for the state's energy boom. While conducting routine surface mapping for economic deposits of clay, state geological survey geologist Fredric F. Mellen observed that the Moodys Branch Formation was present at an abnormally high elevation. The finding was made public in the form of a press release, and oil companies started immediately leasing rights on this apparent domal structure. The Union Producing Company discovered the Tinsley Oil Field on August 29, 1939. The well the company drilled tested at 235 barrels of oil per day from the Woodruff Sand in the Selma Chalk. This discovery of significant oil east of the Mississippi River came less than five months after the press release describing Mellen's discovery.

Of the thirty-three US states with oil and gas production, Mississippi ranks near the middle for oil and slightly lower than that for natural gas. Mississippi's annual oil production peaked in 1970 at 65 million barrels. Since 1950 the state has produced more than 2.3 billion barrels of oil. Annual natural gas production peaked in 1956 at 253 billion cubic feet of gas. Since 1950 the state has produced more than 9.5 trillion cubic feet of natural gas. Carbon dioxide production began in 1978, and since then 3.2 billion cubic feet have been produced.

There are two prominent oil and gas provinces within the state, with production originating from numerous Paleozoic- and Mesozoic-age reservoirs. The first is the Black Warrior Basin of northeastern Mississippi and northwestern Alabama. It comprises a thin layer of Cretaceous- and Tertiary-age strata overlying southwesterly-dipping Paleozoic-age rocks, which are dissected by numerous faults. Production in this basin is primarily from sandstones in the Late Mississippian–age Floyd Shale and Pennsylvanian-age Pottsville Formation. The sands of these sandstones, eroded from the ancestral Appalachians and Ouachitas created during the assembly of Pangaea, were deposited in the basin by rivers flowing from the northeast and southwest. The Mississippian-age Pride Mountain Formation and Hartselle Sandstone in the northeast corner of the state are similar in age to the Floyd Shale, but to date they have not produced any oil or gas. However, the state is currently reviewing the potential of mining heavy oil from the Hartselle at depths as shallow as 50 feet.

There has also been limited production from a Devonian-age chert and the Cambrian-Ordovician-age Knox Dolomite of the Black Warrior Basin. These carbonates were deposited in the Iapetus Ocean on a platform at the edge of the continent. The rocks were fractured during the building of the Appalachians, which gave them the porosity that makes them good reservoir rocks. Although they are capable of trapping oil and gas, the formations have proven only minimally productive. Geologists attribute this to the lack of an organic-rich source rock nearby capable of generating oil and gas.

The second—and largest—oil and gas province is the Mississippi Interior Salt Basin, which has nine regional and three local reservoirs that are productive. With the exception of the Jurassic-age Smackover Limestone and the Cretaceous-age Selma Chalk and James Limestone in the Rodessa Formation, the plays of this province are predominantly in sandstone. Oil or gas has been produced from nearly every formation between the Jurassic-age Norphlet Formation and unnamed Miocene-age sands. The dip of rock units into the Mississippi Embayment provides an interesting dynamic in the salt basin. Rock units exposed in outcrop in northern and eastern Mississippi are buried thousands of feet below the surface in the basin. As a result, rocks in the Cretaceous-age Tuscaloosa and Eutaw Groups and the Paleocene-Eocene-age Wilcox Group that are freshwater aquifers in the northern and eastern parts of the state are oil and gas producers toward the south and west.

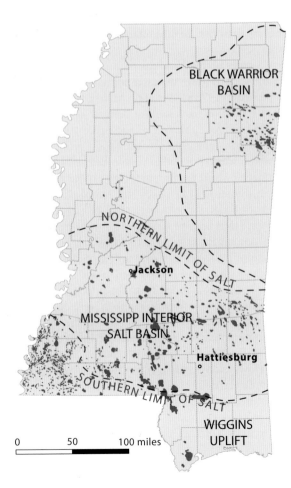

Oil and gas fields (blue) of Mississippi with major basins indicated. (Modified after Thompson 2009.)

NORTHEASTERN MISSISSIPPI

Watch for subtle changes in topography as you drive through northeastern Mississippi. A sudden change in elevation is likely indicative of a change in the underlying geology. If you're traversing a relatively flat area and then start to climb a hill, then you may be crossing from a valley-forming, less resistant unit to a ridge-forming, more resistant unit. The geology and physiography of northeastern Mississippi are the most diverse in the state. The Paleozoic, Mesozoic, and Cenozoic Eras are all represented in outcrop. Although the North Central Hills is the largest province, the Pontotoc Ridge and Paleozoic Bottoms are home to some of the most beautiful vistas and the most impressive geology in the state.

Stream erosion has played a significant role in sculpting the landscape of Mississippi. It is important to note that although streams move a tremendous volume of sediment, they can also deposit sediment and leave behind signs of former river systems. In numerous places in Mississippi you can find relict terrace deposits. A terrace deposit is typically composed of sand or gravel, or both, that was once part of a stream channel. The deposits are generally found on hilltops because they have managed to escape erosion. The river that deposited the material is long gone, but the deposit remains behind as a sign that a river once flowed here. In southern and central Mississippi the terraces are likely composed of the Pliocene-age Citronelle Formation. Along the bluff line in western Mississippi, the terraces are probably Pleistocene-age remnants of the ancestral Mississippi and Ohio Rivers that flowed through the region prior to their entrenchment during the Pleistocene. Still other terraces are found in Jackson and coastal Mississippi that numerous unknown streams formed.

One of the most prominent physiographic features in northeastern Mississippi is the Pontotoc Ridge, a narrow band of resistant rock that extends southward approximately 60 miles from the Tennessee line into Clay County. Hills within the Pontotoc Ridge province rise nearly 800 feet above sea level. Generally, rocks of the Selma Group are composed of chalk. In northern Mississippi, however, the group includes the McNairy Sand and Chiwapa Sandstone Members of the Ripley Formation. The McNairy and Chiwapa are more resistant to erosion than the surrounding rocks and are the primary ridge-forming members of the province. South of Clay County, in Oktibbeha County, all members of the Selma Group transition into less-resistant chalk. Where the sand-rich members of Ripley Formation, deposited close to shore, are replaced by marine chalks deposited farther from shore, the Pontotoc Ridge gives way to the lower-relief Flatwoods and Black Prairie provinces. Pontotoc Ridge serves as the divide between the southwestward-flowing Mississippi and Pearl watersheds and the southerly Tennessee-Tombigbee Waterway watershed.

The Flatwoods physiographic province is also easily distinguishable. It is a long, narrow, arcuate band of low-relief topography that developed on clay of the Porters Creek Formation. The province is immediately west of the Pontotoc Ridge and Black Prairie provinces. Its eastern edge marks the Cretaceous-Tertiary boundary, where Cretaceous-age rocks are in contact with Tertiary-age rocks. This boundary is the division between the Mesozoic and Cenozoic Eras, and it is associated with the meteorite impact at Chicxulub, Mexico, and the mass extinction event that ended the reign of the dinosaurs. The extinction of the dinosaurs about 66 million years ago created opportunities for smaller animals to evolve. This significant change in life on Earth, as represented in the rock record, warranted the establishment of a new era: the Cenozoic is also called the "age of mammals."

LIGNITE

The preservation of plant material in an oxygen-free environment is crucial to the formation of all coal. Swamps and marshes, or wetlands, are ideal locations because they promote plant growth; maintain high water tables, which prevents the plant material from drying out and decaying; have slightly acidic water, which prevents bacteria from consuming the plant matter before it is buried; and are often actively subsiding, which makes room for the continued accumulation of plant material.

If enough wetland plant material is preserved and buried, it eventually forms one of several grades of coal. Burial promotes bacterial decay in an oxygen-free environment and increased temperature, both of which are necessary for coal formation. As plant matter is buried deeper, it may transition through several grades of coal: peat, lignite, subbituminous, and bituminous. If pressures and temperatures are great enough, anthracite, the highest grade of coal, may form. Mississippi's highest grade of coal occurring at the surface is lignite; at depth there are bituminous coals in the Pennsylvanian-age Pottsville Formation.

Coal grades are based on physical properties and energy content. Lignite is soft, easily fractured, and often contains identifiable plant material. It is common in Tertiary-age rocks across the Gulf Coastal Plain, which extends from the Florida panhandle west through coastal Georgia, Alabama, Mississippi, Louisiana, Texas, and into Mexico. (The geology of the coastal plain typically features rocks that are Cretaceous-age and younger.) Lignite-burning power plants are becoming more common in the Gulf states as operators use technology to eliminate many pollutants from the smokestacks.

In Mississippi, coal seams in the Claiborne and Wilcox Groups have the greatest potential for economic exploitation. Lignite seams in the Wilcox can range up to 12 feet thick; one Claiborne seam ranges up to 20 feet thick.

The combined humid, subtropical conditions during Paleocene and Eocene time and subsiding coastal margin of the Mississippi Embayment facilitated wetland formation, plant growth, peat accumulation, peat burial, and eventually lignite formation. The Atchafalaya Basin and Mississippi River Delta in southern Louisiana would be good analogs for Mississippi during Paleocene and Eocene time. There are numerous freshwater wetlands that have developed in abandoned channels and oxbow lakes in the Mississippi Delta today that could someday yield coal, but that's a long way off, and they would be small relative to the older coastal deposits.

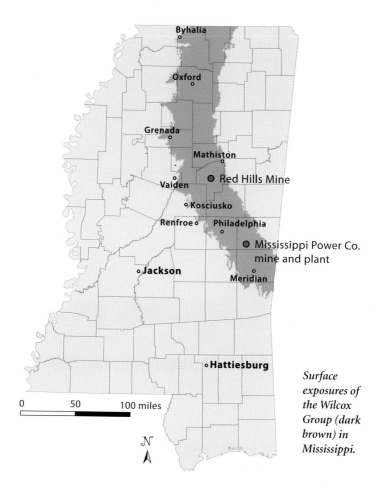

Surface exposures of the Wilcox Group (dark brown) in Mississippi.

A British Thermal Unit (Btu) is the energy required to raise the temperature of one pound of water from 39°F to 40°F. Lignite is on the lower end of the energy spectrum, with Btu values that do not exceed 8,300 Btu/pound (bituminous coal averages 12,000 Btu/pound, anthracite 14,000 Btu/pound). Mississippi lignite from the Wilcox averages 5,570 Btu/pound. Lignite at the Red Hills Mine, in Ackerman, currently averages 5,120 Btu/pound. Because of its lower energy content, lignite typically must be combusted near the mining operation in order to make its production economically feasible. The Red Hills lignite is combusted on-site in the Red Hills Power Plant.

INTERSTATE 20
Jackson—Newton—Meridian—Alabama
112 miles

It just seems appropriate to begin the book's first road log in the capital city of Mississippi, Jackson. Unfortunately, the city of Jackson is underlain by the Eocene-age Yazoo Clay. The Yazoo was deposited well off the coast of Mississippi between 37 and 34 million years ago in approximately 200 feet of water. The clay is famous for two things: well-preserved whale fossils and its ability to shrink and swell with changes in moisture. The whale fossils are a source of wonder, and the Yazoo Clay contains some of the best specimens in the world, but the clay is a source of constant aggravation. Building foundations, road surfaces, and underground pipes are all subject to damage when built on or buried in it.

The Yazoo Clay would not outcrop as extensively in Jackson if the Jackson Dome had not formed here. Near-surface rocks are draped over the top of the dome and sag on the flanks of the structure. As a result, the clay is exposed farther southwest than it would have been had the dome not intruded here. The outcrop patterns for the Oligocene-age Forest Hill Formation and Vicksburg Group behave in a similar fashion. As you traverse the state through Jackson on I-20 you pass over these rocks on both the western and eastern flanks of the Jackson Dome.

Jackson Dome

Within the Mississippi Interior Salt Basin lies one of the most talked-about geologic oddities in Mississippi: a volcanic feature called the Jackson Dome. The top of the dome lies approximately 2,600 feet beneath the surface of the Mississippi State Fairgrounds in Jackson. There have been no indications of resurgent activity.

Age dating done on igneous samples retrieved from oil wells yielded ages ranging from 79 to 69 million years old, making the dome Cretaceous in age. Some of the samples were from lavas, meaning the Jackson Dome did have eruptive events, but the lavas are of a type that would have been very viscous; it's not likely that any of the lava flowed very far from a vent at the surface.

Other volcanic features were active in Mississippi during Cretaceous time. Igneous volcanic rocks found in an oil well in Humphreys County, in west-central Mississippi, were determined to be between 91 and 78 million years old. They are from what is called the Midnight volcano. The suite of volcanic (extruded at the surface) and plutonic (cooled beneath the surface) igneous rocks identified from the Jackson Dome is similar to those of the Magnet Cove complex approximately 10 miles east-southeast

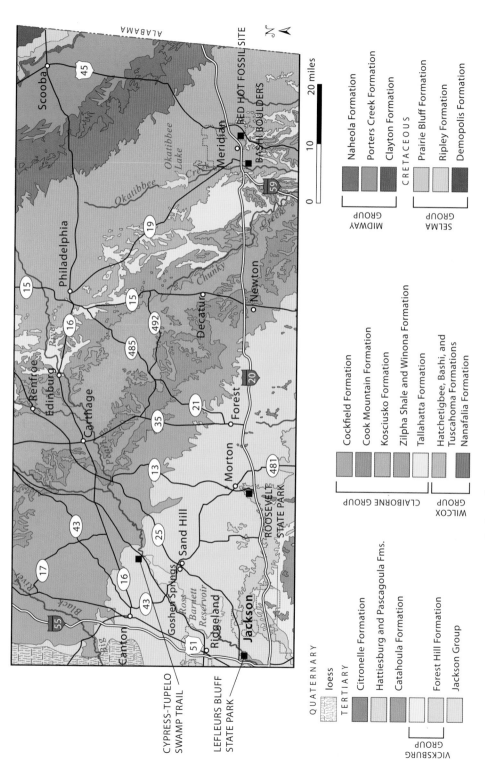

Geology along I-20 between Jackson and the Alabama state line.

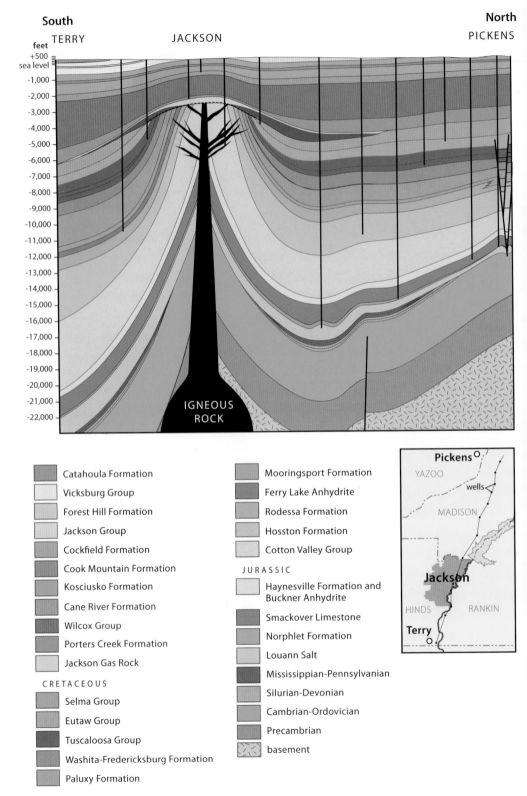

	Catahoula Formation		Mooringsport Formation
	Vicksburg Group		Ferry Lake Anhydrite
	Forest Hill Formation		Rodessa Formation
	Jackson Group		Hosston Formation
	Cockfield Formation		Cotton Valley Group
	Cook Mountain Formation	JURASSIC	
	Kosciusko Formation		Haynesville Formation and Buckner Anhydrite
	Cane River Formation		Smackover Limestone
	Wilcox Group		Norphlet Formation
	Porters Creek Formation		Louann Salt
	Jackson Gas Rock		Mississippian-Pennsylvanian
CRETACEOUS			Silurian-Devonian
	Selma Group		Cambrian-Ordovician
	Eutaw Group		Precambrian
	Tuscaloosa Group		basement
	Washita-Fredericksburg Formation		
	Paluxy Formation		

North-south cross section through the Jackson Dome. Vertical lines are oil and gas wells; information gleaned from these was used to construct the cross section. (Modified after Walkinshaw 2008.)

of Hot Springs, Arkansas. The Magnet Cove rocks are 101 to 88 million years old. Without an active subduction zone or rift there shouldn't be evidence of volcanic activity in the middle of the North American tectonic plate. The arcuate trend of the features toward the southeast from Hot Springs and the progressively younger ages may provide clues as to what happened.

The leading interpretation for the origin of the Jackson Dome is that it is the result of a passing hot spot called the Bermuda hot spot. Hot spots develop where hot portions of the mantle are relatively close to Earth's surface, creating volcanic or igneous activity. The Yellowstone hot spot, responsible for the volcanic activity in Yellowstone National Park, is an example of a famous hot spot. Hot spots stay fixed in place in the Earth's mantle as lithospheric plates pass over them. Like a blowtorch, hot spots heat the lithospheric rocks above them. The North American Plate, of which Mississippi is a part, traveled in a northwesterly direction over the hot spot. This is why the path of the hot spot runs roughly perpendicular to the north-south-trending Mississippi Valley Graben. Indeed, the thinner, fractured crust of the graben is what enabled magma to reach the surface and form volcanic features. With the passing of the hot spot, the crust subsided and formed the Mississippi Embayment, which approximates the path of the Mississippi Valley Graben. Geologists attribute the absence of volcanic rocks southeast of Jackson to the relatively thick crust beneath the Appalachians, which could have prevented magma from reaching the surface.

The igneous rock that makes up the Jackson Dome is a deep-seated feature that behaves like a pier does in the foundation of a house. The rocks deposited on top of the dome are supported, whereas those surrounding the dome have been susceptible to subsidence. As the rocks surrounding the dome sagged, natural gas was able to migrate into the Selma Gas Rock on top of the dome, where it became trapped, forming the Jackson Gas Field. Since its discovery in 1930, the field has produced more than 100 billion cubic feet of natural gas.

Along with the discovery of gas in the Jackson Gas Field came an understanding of the large amounts of heat generated by the Jackson Dome at depth. After failing to produce natural gas with a well they drilled near the current Mississippi Coliseum, the owners of the well generated money in another manner. In the mid-1930s they used warm water produced from the Eocene-age Meridian Sand, at a depth of 2,367 feet, to source a recreational swimming area known as the Crystal Plunge. The name of the attraction, however, was not due to the clarity of the water; it was the owner's name.

A notable recent drilling boom has taken place in deeper formations around the Jackson Dome. However, the target is carbon dioxide used to aid oil recovery efforts in mature oil and gas fields, not natural gas or oil itself. Up to 1 billion cubic feet of carbon dioxide per day has been produced from limestone and sandstone of the Jurassic-age Smackover and Norphlet Formations. The fields are estimated to contain up to 5.6 trillion cubic feet of carbon dioxide. The carbon dioxide is nearly 100 percent pure, and the accumulations are unique in that they are the only such deposits east of the Mississippi River.

From its junction with I-55 at Jackson, I-20 travels on the Pearl River flood-plain and terrace alluvium of Quaternary age until just before mile marker 50. The hill east of MS 468 at exit 49 is the eastern edge of the Pearl floodplain. Between mile markers 50 and 65, I-20 travels through the Southern Pine Hills physiographic province underlain by the Oligocene-age Forest Hill Formation sand and Vicksburg Group limestone and shale. The topography in the South-ern Pine Hills is more rolling than that of the Jackson Prairie because the rocks are slightly more resistant to erosion.

At exit 68 there are numerous cedar trees, which are good indicators of an alkaline soil. Soils are derived from the bedrock on which they rest. Alkaline soils are derived from calcium-rich rocks, such as limestone and chalk, and have pH values above 7, which is neutral. Acid soils have pH values below 7. Pine trees are very well adapted to living in acid soils but not alkaline soils. Although cedar trees can survive in acid soils, they are often outcompeted by pine species. An abundance of cedar trees is an indication that the bedrock is probably a carbonate. The Yazoo Clay and Moodys Branch marl—a clay-rich limestone—in this area contain abundant shell material made of calcite, the primary mineral in limestone.

Between mile markers 65 and 102, I-20 travels across the Jackson Prairie physiographic province, primarily on the Yazoo Clay. However, there are a few intervals where Forest Hill Formation sand and Oligocene-Miocene-age Cata-houla Formation sand outcrop. One of those areas lies between mile markers 75 and 79, where I-20 travels through an outcrop of Forest Hill sand bounded by the Jackson Prairie to the west, north, and east. At Roosevelt State Park, immediately north of exit 77 on MS 13, an overlook at the entrance to the park rests on Pliocene-age Citronelle Formation sand, which lies on top of the For-est Hill Formation. The overlook provides a view across the Jackson Prairie. Springs have formed at the contact between the Forest Hill sand and the under-lying Yazoo Clay; they are visible along several trails in the park.

An occasional pump jack on an oil well may be seen along this segment of road, such as the one on the south side of I-20 at mile marker 67, part of the Morton Field. Production in this area is from several Cretaceous- and Jurassic-age reservoirs.

From mile marker 102, near Lake, to the Alabama state line, I-20 traverses the North Central Hills physiographic province, composed of rocks and sedi-ments of the Jackson and Claiborne Groups (Eocene age) and the Wilcox Group (Paleocene and Eocene age). Between mile marker 102 and the I-59 junction, outcrops belong to the Claiborne Group, including interbedded sand and clay, clay, and marl; none are well exposed. (*Marl* is loosely defined as a limestone that contains abundant clay.) The Tallahatta is the oldest formation in the group and the only unit that stands out in relief. The exposure along this stretch of road was created in the mid-1960s, when I-20 was built. Although fifty years is the blink of an eye in geological time, Mississippi rock units are generally not well cemented and are very susceptible to weathering, erosion, being overgrown by vegetation, or being graded and seeded by the Mississippi

Department of Transportation; the fact that these interbedded siltstone and clay-rich intervals are still visible in road cuts is impressive.

West of here, at the Hinds-Warren county line between Vicksburg and Jackson, these same rocks are approximately 4,400 feet below sea level. That's because on a regional scale, these rocks dip approximately 1° per mile (25 to 30 feet per mile) to the west-southwest due to the Mississippi Embayment. The embayment is a syncline, a large-scale, convex-down fold in which all the strata dip toward the center, or axis, of the structure (see "Mississippi Embayment" on page 36 for more information). Along I-20, the axis is beneath the Hinds-Warren county line. In outcrop you may notice small, localized reversals in the regional dip of the rocks. These small structural features are probably related to near-surface faulting caused by localized subsidence in the Mississippi Interior Salt Basin. The regional dip is very subtle and virtually undetectable along this route.

The rolling topography along this section of I-20 is the result of cuestas, asymmetrical ridges that form due to differential weathering and erosion. Differential weathering and erosion can be caused by a number of factors, such as rock type, degree of cementation, porosity, fracturing, and climate. For example, well-cemented sandstone is more resistant to the forces of weathering and erosion than one that is weakly cemented. As the weakly cemented sandstone weathers and is removed by erosion, the well-cemented sandstone stands out in relief. Along the same lines, a rock that is broken and fractured due to tectonic stress weathers and erodes faster than the same rock that is not fractured. Porosity, or the space between grains in a rock, can also affect a rock or sediment's ability to resist erosion. (Oddly enough, loosely cemented loess can maintain vertical walls when cut because water is adsorbed into the pores and therefore does not run off.) Climate is a big factor in determining relief in the southeastern United States. Carbonate rocks are more easily weathered and eroded in areas with high precipitation because acidic rainfall and groundwater dissolve the rock. In the western United States, where rainfall is not abundant, limestone often stands out in relief and is called a "ridge former." In Mississippi and the southeastern United States, however, sandstones form most of the ridges.

Because of regional dip into the Mississippi Embayment, cuestas in eastern Mississippi have gently dipping western slopes and steep eastern slopes. The Buhrstone and Wilcox Cuestas are located between the Newton-Lauderdale county line and Meridian and between Meridian and the Alabama state line, respectively. The Buhrstone Cuesta is actually composed of several small cuestas. A *buhrstone* is a tough, silica-rich rock that was used to make millstones. The Tallahatta Formation, of early Eocene age, was informally known as the "Buhrstone" or "Siliceous Claiborne." The formation was formally defined in 1897, but the cuesta still bears the original name.

The generally well-cemented Tallahatta is more resistant to weathering and erosion than the underlying Meridian Sand, which is why it stands out in relief in the Buhrstone Cuesta. A similar situation exists within the Wilcox Group, in which the Wilcox Cuesta formed; the Hatchetigbee and Bashi Formation sands are more resistant than those of the underlying Tuscahoma Formation.

The major outcropping between mile markers 124 and 128, just west of the I-59 junction, is composed of the resistant siltstones and sandstones of the Tallahatta Formation, which was deposited in shallow coastal water. The formation contains clay-rich beds as well, but the siltstone and sandstone layers provide much of the relief in this area. The source of the silica cement in the Tallahatta, which makes it so resistant, has generated considerable research over the years. Some investigations suggest it comes from the shells of radiolarians and diatoms, single-celled aquatic organisms that secrete a shell made of silica, whereas others present evidence of volcanic ash as the source. In either case, the high-silica content of the Tallahatta Formation corresponds to a global high in silica values for all rocks deposited around 50 million years ago.

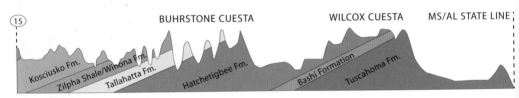

Topographic profile along I-20 showing the geological units that compose the Buhrstone and Wilcox Cuestas. The vertical scale has been greatly exaggerated.

View to the east from near mile marker 130 as I-20 begins to descend the relatively steeper eastern flank of the Buhrstone Cuesta.

Outcrops of the Tallahatta contain numerous *Ophiomorpha* trace fossils, the fossilized burrows of shrimp-like crustaceans. Because only casts of the burrows remain, they are classified as *trace fossils*. The crustaceans burrowed in the Tallahatta when it was still sediment, offshore in water depths probably between 30 and 300 feet. The burrows are preserved because they filled with sediment that was later cemented.

A great 30-foot-tall exposure of the Tallahatta Formation near mile marker 126. The benched outcrop, created during the construction of I-20, is visible in both the eastbound and westbound lanes. The close-up features a weathering profile of more-resistant, silica-rich beds that stand out in relief and less-resistant beds that weather and erode more easily.

Between the junction of I-59 and the Alabama state line, I-20 travels on the Wilcox Group. The Wilcox contains beds of sand, clay, and lignite coal. One of the more interesting formations in the Wilcox is the Bashi Formation. The Bashi, typically 5 feet thick, is composed of a fossil-bearing, glauconitic, white quartz-sand unit bearing boulder-sized, fossil-rich calcareous concretions. The broken fossils were deposited along a beach during Paleocene time. Glauconite is a blue-green mineral that forms from organic matter that is altered in marine environments. The sediment contained in the concretions was better cemented than surrounding material, probably shortly after burial. This selective cementation gives the concretions greater resistance to erosion than other sediments within the same unit. The concretions, known as "Bashi boulders," readily identify this formation in eastern Mississippi and western Alabama.

While traversing the Wilcox Cuesta between mile markers 157 and 165, look for lignite seams. Most Eocene-age formations in Mississippi are lignite bearing, but the seams in the Nanafalia and Tuscahoma Formations are the best developed. These are the seams currently being mined. The Mississippi Power

Relocated Bashi boulders, or large concretions, between 31st Avenue and the frontage road on the south side of exit 152. Each boulder is about 4 feet in diameter. Clams, oysters, and shark teeth are readily visible in the concretions' sediment matrix, evidence of the marine origins of this rock.

Red Hot Fossil Site

The Red Hot fossil site is just south of I-20 and along the eastbound frontage road between exits 153 and 154. Although the truck stop for which it is named is gone, and the surrounding area has been heavily developed, the developer preserved the main fossil locality.

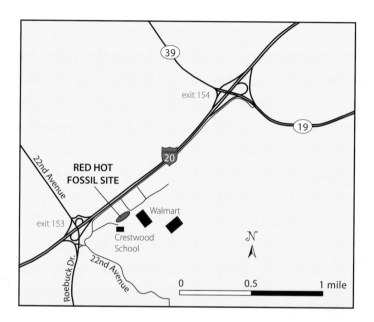

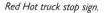

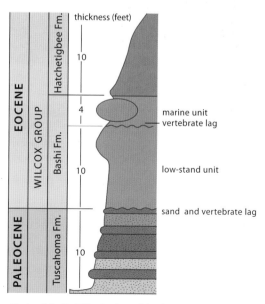

Red Hot truck stop sign.

Rocks of the Red Hot fossil site. (Modified after Ingram 1991.)

Here the Bashi and Tuscahoma Formations of the Wilcox Group have yielded fossils from twenty-four species of mammals; twenty-two species of sharks, rays, skates, and sawfish; eleven species of bony fish; and four species of snakes. Eighteen leaf species that have been identified indicate the region experienced a tropical to subtropical climate when the fossils were accumulating roughly 56 million years ago; the contact between the Bashi and the Tuscahoma is the Paleocene-Eocene boundary. In addition to its faunal diversity, the site is well-known for producing the first Eocene-age mammals east of the Mississippi River, including important specimens of an omomyid primate. Omomyids were the first group of primates found in North America. The specimen from the Red Hot locality probably weighed a pound or so and had gripping hands and feet that featured nails instead of claws.

Red Hot fossil site outcrop immediately west of the western entrance to Walmart. The white arrow indicates the position of a Bashi concretion.

The fossil assemblage is known as the Red Hot fauna, and it rests in lag deposits directly beneath the Bashi Formation (the upper unit of the outcrop), which contains boulders, and the Tuscahoma Formation sand, the fourth and uppermost thin sand layer (approximately 8 inches thick) of the outcrop. Although the outcrop has been somewhat preserved, be aware that it is not easily accessible.

Company plant and mine in Kemper County and the Red Hills Mine in Ackerman (30 and 72 miles to the north-northwest respectively) exploit Eocene-age
lignite. The Kemper County facility will use the lignite to produce a hydrogen-
based synthesis gas to power turbines and generate electricity. The facility will
also remove carbon dioxide, sulfur, nitrogen, and mercury from its emissions.
The lignite from the Red Hills Mine is combusted on-site to generate electricity. The ridges of the Wilcox Cuesta are quite a contrast to the relatively flat
topography to the east toward the Alabama state line. Here much of the weaker
Tuscahoma Formation has been eroded away and is now covered with soil.

*Tuscahoma Formation clay south of I-20 eastbound along the off-ramp at exit 160.
Note the vertical drainage channels, which are characteristic of erosion on steep slopes.*

MISSISSIPPI 16
CANTON—PHILADELPHIA
58 miles
See the map on page 45.

From I-55 at Canton to about 4 miles east of MS 43, MS 16 travels on the Yazoo Clay. The increase in relief about 7 miles east of I-55 indicates a transition to sand and marl of the underlying Moodys Branch Formation, which was deposited in shallow marine water as sea level rose. The gentle, low-rolling terrain of this stretch is very characteristic of the Jackson Group, which makes up the Jackson Prairie.

About 8 miles east of the MS 43 junction, MS 16 transitions onto the Claiborne Group and into the North Hills physiographic province. Sand of the Cockfield Formation outcrops sporadically along this section of highway. The Cockfield is an important aquifer in central Mississippi; the outcrops in this area function as the recharge zone for the aquifer. Precipitation that slowly infiltrates the formation will replenish water withdrawn from wells to the southwest.

An 8-foot-tall outcrop of Moodys Branch Formation sand on the north side of MS 16, about 6.5 miles east of the MS 43 junction. The sands are dark red due to weathering of the iron-rich mineral glauconite. In Jackson, outcrops of the Moodys Branch are an unweathered and dark-gray, fossil-rich, sandy marl.

Cockfield Formation sand at the northeastern corner of the Swamp Road and MS 16 junction, approximately 5 miles east of MS 17. This sand was deposited in a delta environment that was prevalent across central Mississippi during late Eocene time.

Outcrops of the Tallahatta Formation between Edinburg and Philadelphia can be particularly interesting. They often contain excellent crossbeds and *Ophiomorpha* trace fossils, the burrows of shrimp-like crustaceans. The sand beneath the Tallahatta is known as the Meridian Sand. Because the formation dips in the subsurface, water that percolates through the formation here becomes drinking water for homes to the east and southeast. The rise in the road to the east, just past Edinburg and the Pearl River, marks the western edge of the Tallahatta Formation outcrop. When silica-saturated groundwater was exposed to the atmosphere, it precipitated silica cement in the Tallahatta, creating a resistant siltstone and claystone. Volcanic ash was the original source of the silica.

Winona Formation sands are easily differentiated from those of the underlying Tallahatta because they contain a significant amount of glauconite. The glauconitic sands are green when outcrops are fresh but rapidly oxidize to a bright red when exposed to the elements; the Tallahatta is white to tan. Intervals with high iron content often form thin ledges in outcrop. The Winona is approximately 25 feet thick along this route and can often be found capping hills.

Mississippi construction projects usually proceed without having to break up bedrock. However, the improvements of MS 16 through the town of Pearl River, 4.7 miles west of Philadelphia, were an exception. When the Golden

The edge of the well-cemented Tallahatta Formation is responsible for the westward-facing slope on MS 16 just east of the Pearl River.

An outcrop of Winona Formation sand along MS 16, 0.1 mile west of County Road 149 (Black Jack Road). The sand is a remnant of the lowermost Winona, showing up in hillcrests in the Edinburg area just above the Neshoba Sand Member of the Tallahatta Formation.

Moon Hotel and Casino was developed, crews had to extensively excavate a 4-foot-thick Tallahatta sandstone layer to improve the road. The excess material was stockpiled for future building projects on the Mississippi Band of the Choctaw Indian Reservation. Just west of the junction with MS 15, on the outskirts of Philadelphia, MS 16 passes onto older Paleocene-Eocene Hatchetigbee Formation sands of the Wilcox Group.

MISSISSIPPI 19
MERIDIAN—PHILADELPHIA
37 miles
See the map on page 45.

Between Meridian and Philadelphia, MS 19 travels entirely in the North Central Hills physiographic province, predominantly through the Paleocene-Eocene Wilcox Group, which consists of sand, clay, and lignite. There are a couple of exceptions, however, one of which being where MS 19 crosses the Quaternary-age floodplain and terrace alluvium of Okatibbee Creek, just north of Meridian.

The US Army Corps of Engineers created Okatibbee Lake, about 10 miles north of Meridian, by impounding Okatibbee Creek with a 1.2-mile-long earthen dam in 1968. The reservoir is 6 square miles in size and is used for recreation and municipal drinking water. At the Pine Springs Park boat launch, at the southeast end of the lake, 15 feet of Hatchetigbee Formation sand and clay is visible. Scenic routes from the Okatibbee Lake exit at Collinsville travel through the Wilcox Group, with several exposures visible around the lake, including from Okatibbee Water Park on the east side of the lake.

About 1 mile north of the Okatibbee Lake exit, MS 19 passes a few hilltops on which the Eocene-age Meridian Sand of the Claiborne Group is exposed; this is the other exception where MS 19 does not travel on the Wilcox Group. The hills are mostly composed of the sandy Meridian, which is difficult to distinguish from the sands of the underlying Wilcox. The highest hills feature all that remains of the Basic City Member, which is the clay and siltstone portion of the Tallahatta that has not yet eroded away. The hilltop outcrop about 3.5 miles north of the MS 492 junction is the Basic City.

MISSISSIPPI 25
JACKSON—CARTHAGE—RENFROE
60 miles
See the map on page 45.

MS 25 passes through the low-relief Jackson Prairie and North Central Hills physiographic provinces, composed of the Jackson and Claiborne Groups of Eocene age respectively. The Claiborne Group sediments in central Mississippi record changes in sea level. For example, the Cook Mountain

Formation, found in the upper Claiborne Group, transitions from limestone to clay and then is covered over by clay and sand of the advancing Cockfield Formation delta. The Cockfield delta was expanding out into the Gulf as sea level dropped. As sea level began to rise, wave activity reworked shells and material into the Moodys Branch Formation, the lowest unit of the Jackson Group. The Yazoo Clay covered over the Moodys Branch. Floodplain and terrace alluvium of streams occasionally interrupt the clay, sand, marl, and limestone along this route; the most notable is that of the Pearl River at Jackson.

LeFleurs Bluff State Park and Mississippi Museum of Natural Science

The 305-acre LeFleurs Bluff State Park, at the junction of I-55 and MS 25, was named after the French-Canadian explorer Louis LeFleur. In the late 1700s he established a trading post on the banks of the Pearl River, around which a town grew that was eventually named Lefleurs Bluff. The 100-foot-tall bluff the name refers to, composed of the Moodys Branch and Cockfield Formations, is preserved in the park behind the Museum of Natural Science; it's accessible by footpaths. The Moodys Branch is a thin marine sand (10 to 30 feet thick) that was deposited on top of the Cockfield delta as sea level rose. The rise in sea level ultimately resulted in the deposition of the Yazoo Clay.

Basil is a 62-foot-long specimen of Basilosaurus cetoides, one of Mississippi's two state fossils. The other, Zygorhiza kochii, is also a whale. The composite skeleton is 80 percent complete. The skull, forelimbs, and sternum came from Louisiana, and the vertebral column and ribs from Scott County. —Courtesy of Mississippi Museum of Natural Science

The top of the Eocene-age Moodys Branch Formation outcrops within the park in an area formerly known as Fossil Gulch, near the trailhead. This locale served as an alternate type section after the original type section, in the Belhaven neighborhood to the west, was partially filled in. A *type section* is the locality where a formation or member is first described in geological publications. The Moodys also outcrops along Town Creek, south of downtown Jackson and just upstream of the creek's confluence with the Pearl River. The Moodys Branch has yielded the most diverse record of fossils of any formation in Mississippi. There have been 204 species of mollusks identified from the Town Creek locality. Snails, clams, scallops, octopuses, and squids are the most common mollusks that have been found.

The Mississippi Museum of Natural Science within the park is home to Basil, the skeletal remains of the whale *Basilosaurus cetoides*. Basil's skeleton is actually a composite of fossils recovered from sites in Scott County and Red River Parish, Louisiana. During the late Eocene whales thrived in an inland extension of the Gulf of Mexico known as the Jackson Sea. The Yazoo Clay, deposited on the ocean floor, was the final resting place for numerous whale and shark fossils, including Basil. The museum also houses a great display of the state's fossils by geologic age, including those from the Town Creek locality.

About 2 miles east of I-55, at the southeastern corner of the junction of MS 25 and Ridgewood Drive, there is a transformer station. A marker on the wall indicates the high-water mark of the 1979 Easter flood. MS 25 was inundated from I-55 to the edge of the floodplain, 3 miles to the east at the Airport Road (MS 475) junction. Most of the development along MS 25 has taken place since the 1979 flood and was accomplished by raising much of the land to road level.

Much of the sand and gravel fill used to raise structures out of the floodplain was mined from terraces in the area that were deposited by ancestral rivers. Some of the terraces that remain at the highest elevations in central Mississippi bear minerals and rocks suggesting the sand and gravel was sourced from the Appalachians via the ancestral Tennessee and Ohio Rivers. Many of these deposits are isolated on hilltops and have not been fully investigated because they are difficult and expensive to date. In the Jackson metropolitan area the medium-grained sand of these terraces occur at higher elevations. They have not been precisely dated, but they were deposited after the top of the underlying Yazoo Clay had begun to erode.

From the Carthage city limits to Renfroe, MS 25 traverses the Kosciusko Formation, which is composed primarily of crossbedded delta sand with lesser amounts of silt and clay. Ancient rivers deposited the Kosciusko as they flowed into the Gulf of Mexico. Watch for numerous exposures between the MS 13

An outcrop of an undated terrace at the northwestern corner of the MS 25 and Manship Road junction, about 8 miles east of I-55. The white grains in the close-up are granules of chert.

The Yazoo Clay outcrop at Goshen Springs contains gypsum crystals (reflecting sunlight) up to 1 inch long that form in fractures in the clay.

and MS 19 junctions; the eroded embankment on the southeastern side of MS 25, about 3 miles past the MS 16 junction, is the Kosciusko Formation. Sand intervals deeper within the Kosciusko serve as aquifers in much of the Jackson area. Drainage ditches at road level have eroded through iron-cemented sandstone, revealing crossbedding and *Ophiomorpha* burrows, the fossilized burrows of shrimp-like crustaceans that tunneled through the sand when it was near the shoreline. Crossbedding is a type of structure that indicates sediment was transported by moving water or wind. Crossbeds form in various depositional environments, including those of rivers, beaches, shallow or deep oceans, and deserts. Crossbeds feature bedding planes that intersect at angles to one another.

As MS 25 approaches the community of Renfroe, it crosses relatively flat terrain composed of Quaternary-age floodplain and terrace alluvium for about 1 mile before reaching Lobutcha Creek, which deposited the alluvium.

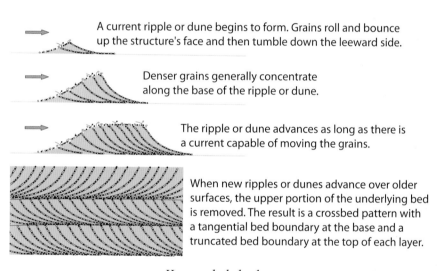

A current ripple or dune begins to form. Grains roll and bounce up the structure's face and then tumble down the leeward side.

Denser grains generally concentrate along the base of the ripple or dune.

The ripple or dune advances as long as there is a current capable of moving the grains.

When new ripples or dunes advance over older surfaces, the upper portion of the underlying bed is removed. The result is a crossbed pattern with a tangential bed boundary at the base and a truncated bed boundary at the top of each layer.

How crossbeds develop.

Goshen Springs

In Goshen Springs there's an interesting outcrop of Yazoo Clay along the northbound entrance ramp to MS 25 from MS 471. The clay contains gypsum crystals that form when calcium (derived from shells) and sulfur (derived from pyrite, or fool's gold) and oxygen combine. Gypsum is the primary ingredient in plaster of paris and drywall. The clay is very slippery and sticky when wet, so be prepared for extensive cleanup should you venture off the road to view the outcrop when the ground is wet. The gypsum crystals will not be present immediately after significant rainfall because they dissolve in the rain.

INTERSTATE 55

CANTON—TILLATOBA

100 miles

Between Canton and Tillatoba, I-55 predominantly passes through the North Central Hills physiographic province. There are few exposures of the formations that underlie this region.

From Canton, I-55 briefly passes over the low-relief Yazoo Clay of the Jackson Group before encountering the Claiborne Group, both of Eocene age. For the most part, the Claiborne sediments are sandy and thus lead to hilly relief because they are resistant to erosion.

Between mile markers 127 and 132, the highway crosses the terrace and floodplain alluvium of the Big Black River. Note the relatively flat topography of the current floodplain and the elevation differences between it and the nearby terraces, the river's former floodplains, such as at mile marker 130. If age dates were available for the terraces, geologists would be able to determine the river's rate of down-cutting. Another notably low-relief stretch occurs between exits 206 (MS 8) and 211 (MS 7), where I-55 crosses the alluvium of the Yalobusha River. Quaternary-age loess—windblown silt deposits—caps the hilltops near exit 220 (MS 330) at Tillatoba, but it is too thin to easily recognize.

MISSISSIPPI 35

FOREST—VAIDEN

74 miles

In northeast Mississippi, MS 35 travels through the Jackson Prairie and North Central Hills physiographic provinces. From the I-20 junction through Forest the highway crosses the Yazoo Clay of the Jackson Group, but from just north of MS 21 to I-55, at Vaiden, the rest of the route passes over sand and clay of the Eocene-age Claiborne Group. There are few outcrops.

During construction of the Clearview Landfill, 10 miles southeast of Forest, workers uncovered one of the most complete skeletons of *Basilosaurus* ever found in Mississippi. When fossils of this creature were initially discovered in the nineteenth century, early researchers regarded the creature as a reptile, hence its common name "king lizard." *Basilosaurus* is actually an early whale that swam the warm seas of late Eocene time. Curiously, this whale had small legs that may have been a residual feature from terrestrial ancestors. The Clearview skeleton measured more than 61 feet in length, from the skull to the last vertebra. These remains and those of another skeleton, found in Louisiana, were combined to create the composite found hanging in the Mississippi Museum of Natural Science in Jackson.

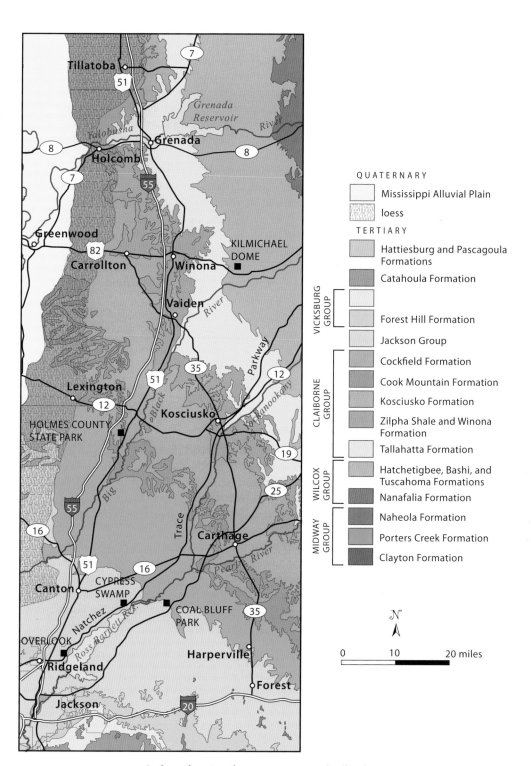

QUATERNARY

Mississippi Alluvial Plain

loess

TERTIARY

Hattiesburg and Pascagoula Formations

Catahoula Formation

VICKSBURG GROUP
Forest Hill Formation

Jackson Group

CLAIBORNE GROUP
Cockfield Formation

Cook Mountain Formation

Kosciusko Formation

Zilpha Shale and Winona Formation

Tallahatta Formation

WILCOX GROUP
Hatchetigbee, Bashi, and Tuscahoma Formations

Nanafalia Formation

MIDWAY GROUP
Naheola Formation

Porters Creek Formation

Clayton Formation

N

0   10   20 miles

Geology along I-55 between Canton and Tillatoba.

The hills just north of Forest and then again between the floodplains of the Pearl and Yockanookany Rivers are areas that smaller tributary streams are still actively eroding. MS 35 travels up and down over countless ravines and divides. Some of the hills are interrupted by floodplain and terrace alluvium of several streams, most notably that of the Pearl, beginning near Carthage; the Yocka-nookany, between MS 19 and the Natchez Trace Parkway; and the Big Black River, just southeast of Vaiden.

The Big Black River's floodplain is currently about 100 feet lower than both the Pearl and Yockanookany floodplains. At some locations near Kosciusko, tributary streams of the Big Black and the Yockanookany watersheds are less than 1 mile apart, separated by a divide. In time, as the Big Black watershed expands, it will capture the Pearl watershed (including the Yockanookany) somewhere north of Jackson. When one river captures another river's water-shed it's called stream piracy. The red sand outcropping on the east side of the road about 0.5 mile past the US 51 junction is the Winona Formation.

NATCHEZ TRACE PARKWAY
RIDGELAND—KOSCIUSKO
60 miles
See the map on page 65.

The Natchez Trace Parkway is a National Park road that closely follows the course of the original Trace, which runs from Natchez to Nashville, Tennessee. The Trace originally was a Native American trail first mapped by the French in 1733. It became an important wilderness road, the most heavily traveled in the "Old Southwest," until steamboats provided a faster and safer mode of travel in the early 1800s.

Between Ridgeland and MS 16, from south to north, the Natchez Trace Park-way traverses the Jackson Prairie and the North Central Hills physiographic provinces, which are underlain by the Eocene-age Jackson and Claiborne Groups and the Eocene-Paleocene Wilcox Group. Unfortunately, there are no outcrops along this stretch, but there are two important features of the land-scape worth pointing out: Reservoir Overlook at milepost 105.6 and Cypress Swamp at milepost 122.

The gentle hills just north of Cypress Swamp indicate that the bedrock has changed from the Yazoo Clay to Cockfield Formation sands of the Claiborne Group. Between the swamp and mile marker 135 at MS 16, the Trace follows the edge of the western floodplain of the Pearl and Yockanookany Rivers. For the most part, the Claiborne Group sediments are sandy, creating hilly relief due to their resistance to erosion. Exceptions include relatively low-relief stretches through the Zilpha Shale and Tallahatta Formation clay, between mile markers 171 and 179.

Ross Barnett Reservoir and Cypress Swamp

The Ross Barnett Reservoir project was commissioned in 1960 to provide economic development, drinking water, and recreational opportunities in central Mississippi. The 52-square-mile lake was completed in 1965 by impounding the Pearl River. It has an average depth of 12 feet and features 105 miles of shoreline, including the 3.5-mile-long dam and spillway. The location of the reservoir has much to do with the Yazoo Clay, which serves as the lake bottom. Because the clay is nearly impermeable, it is an excellent natural liner; nothing had to be added to the clay to make it hold water. Although the Ross Barnett was not built to provide flood control, following the 1979 Easter flood in Jackson the Pearl River Valley Water Supply District elected to manage pool levels to help mitigate downstream flooding potential. Winter and spring levels are typically 1.5 feet lower than summer levels.

Looking east across the Ross Barnett Reservoir.

Cypress Swamp, at the northeastern tip of the Ross Barnett Reservoir and just south of the boundary separating the Jackson Prairie and North Central Hills, is worth viewing and hiking through. The 0.5-mile-long trail with boardwalks offers an up-close look at the advanced stages of an oxbow lake produced when the Pearl River abandoned its channel. Rivers play a major role in sculpting the landscape in Mississippi, but their channels are geologically short-lived features. Rivers flowing over landscapes with very slight changes in elevation, such as those in Mississippi, start to develop very pronounced curves known as meanders. When a river cuts a shorter route through the meander, the meander is eventually cut off from the main channel and abandoned. Once abandoned, the channel begins to fill in with sediment. The meander at Cypress Swamp is in the late stages of its life, having been filled in with a lot of sediment. Although the age of the meander hasn't been determined, the mature cypress and tupelo trees could be hundreds of years old. This swamp is an excellent example of one of Mississippi's classic freshwater wetland environments.

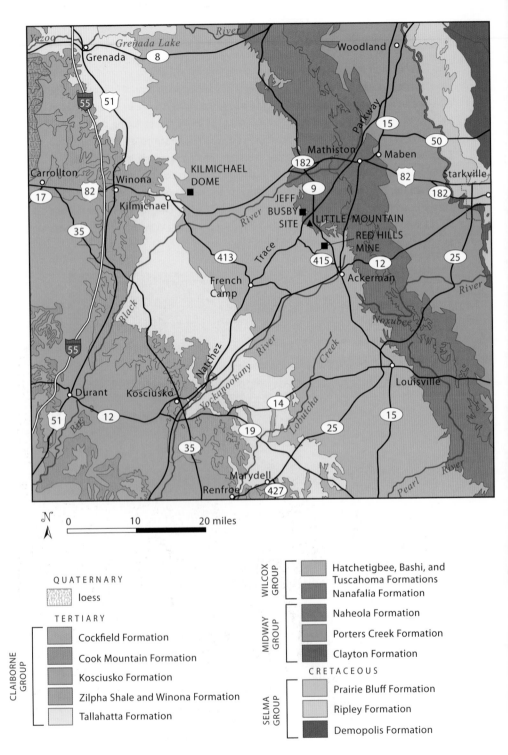

River
Grenada Lake
Grenada
8
Woodland
55
51
Parkway
15
50
Mathiston
182
Maben
Starkville
KILMICHAEL
DOME
Carrollton
Winona
9
82
17
Kilmichael
River
JEFF
BUSBY
SITE
LITTLE MOUNTAIN
182
35
RED HILLS
MINE
413
Trace
25
French
Camp
415
12
Ackerman
River
Black
55
Noxubee
Creek
Natchez
River
Louisville
Durant
Kosciusko
Yockanookany
14
51
12
River
Lobutcha
15
19
25
35
Marydell
Renfroe
427
Pearl
River

N
0   10   20 miles

QUATERNARY
loess

TERTIARY

CLAIBORNE
GROUP

Cockfield Formation

Cook Mountain Formation

Kosciusko Formation

Zilpha Shale and Winona Formation

Tallahatta Formation

WILCOX
GROUP

Hatchetigbee, Bashi, and
Tuscahoma Formations

Nanafalia Formation

MIDWAY
GROUP

Naheola Formation

Porters Creek Formation

Clayton Formation

CRETACEOUS

SELMA
GROUP

Prairie Bluff Formation

Ripley Formation

Demopolis Formation

Geology along US 82 between Carrollton and Starkville.

US 82
CARROLLTON—STARKVILLE
70 miles

From Carrollton to about 1 mile east of the junction with MS 413 at Kilmichael, US 82 travels on the Eocene-age Claiborne Group. Red Kosciusko sand is exposed at the I-55 exit and the hilltop 1 mile from the exit ramp. Along the 2-mile stretch east of US 51, the bright-red exposures spotted through vegetation or in eroded areas are the Winona Formation sand. The Winona is well known for its bright-red color. The green, iron-bearing mineral glauconite, which formed on a shallow seafloor as sediment, oxidizes to red when exposed to the atmosphere. In many Winona outcrops, resistant ledges of iron-cemented sandstone form, and fossils, the original mineral matter of which has been replaced by iron minerals, can be found. About 4 miles east of I-55, US 82 crosses into the sands and silty clays of the Tallahatta Formation.

About 1 mile east of the MS 413 junction at Kilmichael, US 82 travels just to the south of a noted geologic feature known as the Kilmichael Dome,

A 10-foot-tall exposure of Kosciusko Formation sand on the north side of US 82 behind the Winona Church of God Family Life Center, 1.5 miles east of the I-55 junction. (There is also a westward-facing mine on the south side of the road with good Kosciusko exposure.) This middle Eocene sand was deposited on the shallow continental shelf during a regressive event.

An outcrop of Wilcox Group silt and clay, possibly the Hatchetigbee Formation, on the south side of US 82, 2 miles east of the MS 182 junction at Kilmichael. The silt and clay of the Hatchetigbee were deposited in a delta environment.

Kilmichael Crater, or Kilmichael Meteor Crater. The 5-mile-diameter feature is ringed with faults. At the surface, the Midway Group is exposed in the center and surrounded by the Wilcox Group and Tallahatta and Winona Formations. This outcrop pattern identifies the feature as a dome structure. At depth, the structure features a central uplifted area, 1 mile in diameter, composed of rocks that are nearly 1,500 feet higher than they should be. The central area is surrounded by a circular depression in which rocks are nearly 600 feet lower than they normally would be. This subsurface structure is characteristic of other documented impact structures around the world.

Geologists have studied the Kilmichael Dome since its discovery in the 1930s. Geophysical research in the early 1980s suggested it was a meteorite impact structure, and this remains the leading theory explaining its origins. When meteorites strike Earth they have enough energy to cause changes to mineral structure. Opponents of the impact theory cite the lack of "shocked" mineral grains in the dome's rocks as evidence against meteorite impact. Recent research, however, documented rubble zones and broken rock at depth, which is consistent with an oceanic impact site. If the impact occurred under water, shocked minerals, such as quartz, would not necessarily have been produced.

Several other hypotheses for the dome's origins have been suggested over the years, including that it is a buried volcano, a salt dome, or a flower structure (a complexly faulted zone created by horizontal strain in the crust). Both the volcano and salt dome hypotheses have been eliminated, but the meteorite impact and flower structure hypotheses remain viable. A proposal was made to designate the area as a National Natural Landmark, but the region has not been recognized.

From about 1 mile east of the junction with MS 413 at Kilmichael to about 5 miles past MS 9, US 82 travels through the Wilcox Group of Paleocene-Eocene age. The Wilcox consists of sand, clay, and lignite beds deposited by a large delta system that was located in central Mississippi. These beds are exposed on the north side of US 82 about 6 miles east of the junction with MS 182 at Kilmichael. For about 1 mile, starting just west of the junction with Natchez Trace Parkway at Mathiston, US 82 crosses the Quaternary-age floodplain and terrace alluvium of the Big Black River.

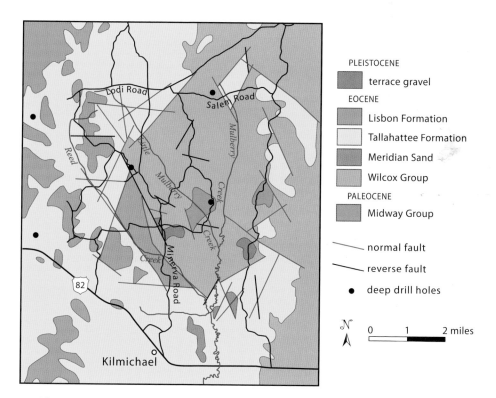

The terrain and forest cover make it extremely difficult to examine the surface geology of the Kilmichael Dome. High precipitation and the growth of vegetation destroy outcrops exposed during excavation and road construction. (Modified after Robertson and Butler, 1982.)

A 5-foot-tall exposure of the Naheola Formation on the south side of US 82 immediately west of the junction with the Natchez Trace. This sand was deposited in a delta environment.

At the junction with the Natchez Trace, US 82 crosses the contact between the Nanafalia Formation of the Wilcox Group and the sandy Naheola Formation of the Midway Group, both of Paleocene age. The Naheola Formation is the easternmost geologic unit of the North Central Hills province. The outcrop pattern of the Naheola defines an arcuate, asymmetrical ridge, or cuesta, that trends toward the southeast. The ridge forms the boundary between the Tennessee-Tombigbee watershed to the east and the Big Black to the west. At Mathiston the eastern slope of the cuesta is steeper because tributary streams of the Tennessee-Tombigbee River erode into the relatively soft Porters Creek Formation clay underlying the Naheola. The more resistant sands in the Naheola provide for more gentle slopes to the west, into the Big Black basin.

From about 4 miles east of MS 15 to 4 miles east of the MS 25 junction, at Starkville, the road mostly travels on marine clay of the Porters Creek Formation. The clay was deposited in deeper water on a coastal shelf in the Mississippi Embayment as sea level gradually lowered from the maximum high of Paleocene time. The gently rolling terrain is characteristic of the Flatwoods physiographic province.

MISSISSIPPI 15
PHILADELPHIA—WOODLAND
104 miles
See the map on page 68.

From Philadelphia to roughly 8 miles north of Ackerman, MS 15 travels through sand, clay, and lignite of the Wilcox Group of Paleocene-Eocene age. From there to Maben, the road travels on fine-grained sand and clay of the Paleocene-age Naheola Formation, and from Maben to Woodland on the Porters Creek Formation, also composed of sand and clay. The notable relief of this portion of the North Central Hills physiographic province is the result of these primarily sandy formations that resist erosion.

A 20-foot-tall exposure of Tuscahoma Formation sand on the east side of MS 15, about 5 miles north of the MS 12 junction at Ackerman. Streams that were part of a large delta system of the Paleocene coastline deposited the Tuscahoma, part of the Wilcox Group.

MISSISSIPPI 25
RENFROE—STARKVILLE
61 miles
See the map on page 68.

Out of Renfroe MS 25 traverses primarily the Zilpha Shale and Winona Formation, with minor outcrops of the Kosciusko and Tallahatta Formations. Look for a large, 10-to-12-foot-tall outcrop of red Winona Formation sand on the west side of MS 25 about 1 mile north of Mars Hill Road, at Marydell (about 1 mile south of MS 427). The Winona, deposited in a continental shelf environment during Eocene time, is fossil rich here. Iron oxide produced by the weathering of glauconite has replaced the original material of many of the fossil mollusks (clams and snails), leaving molds and casts.

A few miles north of the MS 19 junction, the road passes onto sand, shale, and lignite of the Paleocene-Eocene Wilcox Group, deposited in the delta environment of a coastal plain. Approximately 9 miles north of the junction with MS 15N, MS 25 crosses from the Central Hills physiographic province into the Flatwoods province, which is underlain by clay of the Paleocene-age Porters Creek Formation. The clay was deposited in deeper water of the Mississippi Embayment during and after the maximum sea level rise, or transgression, of the Paleocene shoreline. A ridge composed of the Naheola Formation, a sandy deposit laid down by streams as sea level dropped, separates the two provinces.

Wilcox Group clay on the west side of MS 25, about 0.2 mile north of Midway Road.

From the ridge the road descends into the Noxubee River valley and on into Starkville, which rests on the eastern edge of the Flatwoods province.

The road crosses notable floodplain and terrace alluvium of the Noxubee River (1.5 miles) and Cypress Creek (1 mile). The Noxubee River deposits are less than 0.5 mile north of the Winston-Oktibbeha county line. Cypress Creek is about 2 miles north of this county line.

For about 1 mile north of the junction with MS 12, MS 25 travels through the Clayton Formation, which is primarily a clay unit at this locality, before passing onto the Prairie Bluff Formation, a chalk of the Cretaceous-age Selma Group. It passes through this chalk to the US 82 junction. The Prairie Bluff is fossil rich in places, and it makes up the portion of the Black Prairie physiographic province that this route passes through.

The Cretaceous-Tertiary boundary *is* visible in Mississippi, but high rates of erosion, rapid vegetative growth, and the Mississippi Department of Transportation quickly cover many good exposures. However, this boundary can be found about 1 mile north of the MS 12 junction on the west side of MS 25. Here, the Clayton Formation clay (of Tertiary age) rests on top of the Prairie Bluff Formation (of Cretaceous age). This boundary marks the major extinction event that occurred 66 million years ago, which wiped out the dinosaurs and many other species.

A 30-foot-tall exposure of the Prairie Bluff chalk in a drainage east of MS 25, approximately 1 mile north of the junction with MS 12.

NATCHEZ TRACE PARKWAY
Kosciusko—Mathiston
See the map on page 68.

Between Kosciusko and Mathiston, the Natchez Trace Parkway crosses the Claiborne, Wilcox, and Midway Groups. The sand, silt, and clay of the groups' formations record multiple transgressions and regressions of sea level during Paleocene and Eocene time. The sands were deposited in shallow water and terrestrial environments, the silt and clay in deeper water of the continental shelf. As you travel northeast you travel onto older rocks, thus further back in time. Very few outcrops occur along this route.

Red Hills Mine and Power Plant

Mississippi officially became a coal mining state in 1999, when the 5,800-acre Red Hills Mine began operating in Ackerman, in Choctaw County. It is the first lignite mine in the state. The mine, operated by the Mississippi Lignite Mining Company, produces an average of 3.2 million tons of lignite annually. Of the six currently mined seams, five are situated in the Grampian Hills Member of the Nanafalia Formation, which belongs to the Wilcox Group. The remaining seam is in the overlying Tuscahoma Formation. There is approximately 5 feet of waste rock to every foot of lignite in the 140-foot mined interval of the Grampian Hills Member. Because lignite does not produce a lot of energy when burned, a power plant was built on-site to make the

Aerial perspective of the operations at Red Hills Mine showing multiple lignite seams.
—*Courtesy of North American Coal*

mining economically feasible. The Mississippi Power Company, which operates the power plant, has permitted a second lignite mine in Kemper County, which could mine up to 4.1 million tons of lignite per year.

The mine and power plant are located southeast of Little Mountain. Take MS 9 south for about 11 miles and turn right on Pensacola Road to enter the industrial park, power plant, and mine. If you inquire ahead of time, field trips into the mine are available. There are a few road cuts along MS 9 that expose sand, silt, and clay of the Wilcox Group.

The Wilcox, which the Trace starts crossing southwest of French Camp, consists of beds of sand, clay, and lignite coal deposited on a coastal plain in delta environments. Note the relatively high relief, which is typical of the Wilcox because it is rich in sands that resist erosion. The Jeff Busby Site (mile marker 193) is situated on the Wilcox Group. Nearby Little Mountain reaches one of the highest elevations (594 feet) in Mississippi. North of Little Mountain, the Trace traverses the Wilcox Group almost all the way to Mathiston.

INTERSTATE 22/US 78
TENNESSEE—NEW ALBANY
69 miles

I-22/US 78 passes through the Loess Hills physiographic province from the Tennessee state line to Byhalia. The loess caps Pleistocene-age pre-loess alluvial gravel and Kosciusko Formation sand. The loess is thin to nonexistent by the time you reach exit 14. From near exit 14 at Byhalia to just east of mile marker 38, I-22/US 78 travels on the Kosciusko and Tallahatta Formations of the Eocene-age Claiborne Group. From there to just west of mile marker 55, near Myrtle, it travels through the Eocene-Paleocene Wilcox Group, which consists of beds of sand, clay, and lignite coal. Both of these groups are sandy and resistant to erosion, thus they form hilly terrain, which is part of the North Central Hills physiographic province.

Potts Camp was the site of one of the state's first successful pig-iron furnaces. In 1913 Memphis Mining and Manufacturing produced 120 tons of pig iron from Wilcox Group sand. The iron was in the form of iron carbonate, or siderite, not iron oxide or iron hydroxide as in most large-scale mining operations. Pig iron is a form of iron that contains higher amounts of carbon and other impurities than other types of iron, making it brittle. Pig iron was the feedstock for blacksmiths making wrought iron products. In the 1960s, state geologists documenting mineral reserves in the state investigated the potential of the Porters Creek Formation, Naheola Formation, and Winona Formation to yield commercial iron reserves. Before that, the Kilmichael Mining Company had mined these formations and shipped siderite ore to Birmingham, Alabama,

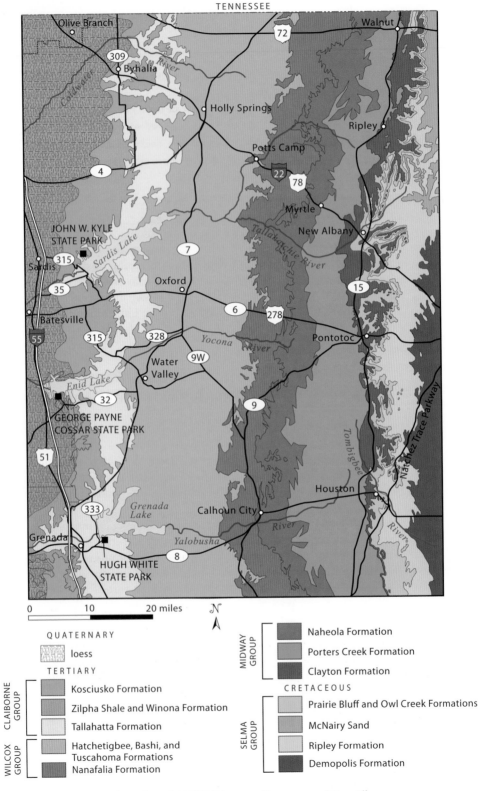

TENNESSEE

Geology along I-22/US 78 between Tennessee and New Albany.

QUATERNARY

loess

TERTIARY

CLAIBORNE GROUP
- Kosciusko Formation
- Zilpha Shale and Winona Formation
- Tallahatta Formation

WILCOX GROUP
- Hatchetigbee, Bashi, and Tuscahoma Formations
- Nanafalia Formation

MIDWAY GROUP
- Naheola Formation
- Porters Creek Formation
- Clayton Formation

CRETACEOUS

SELMA GROUP
- Prairie Bluff and Owl Creek Formations
- McNairy Sand
- Ripley Formation
- Demopolis Formation

until the 1950s, but there has not been any commercial iron mining since that time, leading one to conclude that the potential for the formations to be economically viable is minimal.

From just west of Myrtle to just west of mile marker 60, I-22/US 78 travels on the clay of the Porters Creek Formation. Because clay erodes easily, this formation creates low-relief topography, which makes up the Flatwoods physiographic province. From exit 64 at New Albany to mile marker 69, I-22/US 78 travels through sandy limestone and sand of the Clayton Formation. The Clayton and Porters Creek Formations both belong to the Midway Group; they were deposited in shallow water during the first marine transgression into the Mississippi Embayment during Paleocene time. The Clayton Formation is the oldest Paleocene-age rock unit exposed in Mississippi, and it often contains fossils of Cretaceous age that were worked loose from the underlying Owl Creek sands and Prairie Bluff chalk as the sea rose and spread across the eroded Cretaceous landscape.

During the Late Cretaceous, the Owl Creek and Prairie Bluff Formations were deposited at the same time but in different environments. North of New Albany the Owl Creek Formation is sandy because it was closer to a source of sediment runoff from the coast. The Prairie Bluff Formation chalk occurs south of New Albany and was deposited farther offshore, away from the influence of runoff. New Albany has been developed on three formations: the Ripley, Prairie Bluff, and Clayton. The Ripley is composed of fossil-rich sand deposited in shallow coastal waters in the Late Cretaceous.

The most famous member of the Ripley Formation is known as the Coon Creek Tongue. The Coon Creek is the sandy bottom layer of the Ripley that contains abundant mollusk and crab fossils. Every time the Ripley Formation is excavated in the New Albany area, the Coon Creek garners interest due to the fossils that are unearthed.

US 278/MISSISSIPPI 6
BATESVILLE—OXFORD—PONTOTOC
53 miles
See the map on page 78.

From I-55 near Batesville to 6 miles beyond the Panola-Lafayette county line, US 278/MS 6 travels through the Kosciusko and Tallahatta Formations of the Eocene-age Claiborne Group. Loess occasionally caps the formations for the first part of this drive, in the Panola County portion. This windblown silt was deposited by westerly winds at the end of Pleistocene time.

US 278/MS 6 transitions onto the Wilcox Group of Paleocene-Eocene age. The Wilcox consists of beds of sand, clay, and lignite coal deposited in a variety of delta environments, ranging from floodplains to stream channels to wetlands to offshore. The predominantly sandy units of the Claiborne and Wilcox Groups make up the North Central Hills physiographic province. Some layers have greater resistance to weathering and erosion due to the degree to which the sand was cemented. Layers that are less resistant have eroded into valleys and flat land.

About 17 miles east of MS 7 in Oxford, US 278/MS 6 first crosses onto sand of the Paleocene-age Naheola Formation and then clay of the Porters Creek Formation, which it crosses nearly all the way to Pontotoc. The clay was deposited in deep water as sea level dropped in the Mississippi Embayment following the maximum transgression of Paleocene time. The gently rolling terrain of the clay makes up the Flatwoods physiographic province. Clay units are generally not well cemented, are composed of smaller particles, and are more uniform in composition than sand-rich units. All of these factors result in more easily weathered material that does not produce sudden changes in elevation. About 1 mile west of the MS 15 junction, US 278/MS 6 crosses through Paleocene-age Clayton Formation sand.

University of Mississippi

At the University of Mississippi (Ole Miss) in Oxford, State Geologist Eugene W. Hilgard measured one of the first sections in Mississippi in the mid-1860s. When geologists "measure" an outcrop or road cut, they describe the rock type and unique features, including fossils, and measure the thickness of the exposure. This work is especially important in places such as Mississippi, where erosion and vegetation quickly destroy exposures. The former railroad cut that Hilgard measured is now on Gertrude Ford Boulevard at the overpass for University Avenue. In Hilgard's time, the Mississippi Geological Survey was located on the Ole Miss campus; it was housed in Ventress Hall until 1963. Hilgard went on to become the "father" of soil science.

Hilgard Cut as it appears today. The exposure is approximately 24 feet tall. —Courtesy of Catherine Henry

MISSISSIPPI 8
GRENADA—NATCHEZ TRACE
58 miles
See the map on page 78.

Between Grenada and the Natchez Trace, MS 8 travels through two physiographic provinces: the North Central Hills and a portion of the Pontotoc Ridge. The Tallahatta Formation, Wilcox Group, and Naheola Formation make up the North Central Hills. From Grenada to about 5 miles east of the exit to Hugh White State Park, MS 8 travels through the Tallahatta, except where it crosses the Quaternary-age floodplain and terrace alluvium of the Batupan Bogue. Silt primarily composes the upper portion of the Tallahatta, though in places it outcrops as siltstone. Some Tallahatta outcrops contain ironstone and ironstone ledges, which stand out in relief due to this rock's increased resistance to weathering and erosion. Ironstone is common in Mississippi. When iron is leached from iron-bearing minerals such as glauconite, it can be precipitated around sand grains as cement.

There are exposures of the Tallahatta along the shore of Grenada Lake in Hugh White State Park. Grenada Lake, most accessible off MS 8, was created in 1954 to aid with flood control in the Yazoo Basin, to the southwest. A 2.6-mile-long earthen dam impounds the Yalobusha and Skuna Rivers, creating the 55-square-mile lake with 54 miles of shoreline. The Meridian Sand, the lowest member of the Tallahatta, outcrops on the western shore in several places at water level. The Meridian was deposited in shallow water, where sand grains were reworked by wave activity during Eocene time. Reworking results in rounded grains that are similarly sized and a deposit with low clay content, properties that help enhance the porosity of the Meridian and make it a good aquifer in the subsurface.

Mica-rich shale and claystone found at the park are informally named "paper shale" because they can be split into paper-thin slices. The paper shale is part of the Basic City Member and is the oldest unit of the Tallahatta Formation. Geologists interpret the shale as having been deposited in a coastal marsh setting. When split, this shale often reveals plant fossils. Up to twenty-seven plant species have been collected at Hugh White State Park.

About 5 miles east of the Hugh White State Park exit, MS 8 passes onto the Wilcox Group, which consists of sands, clays, and lignite coal. It passes through the Wilcox to the MS 9S junction. Between MS 9S and MS 9N, in Calhoun City, MS 8 travels on Quaternary-age floodplain and terrace alluvium of the Yalobusha River.

The contact between the Nanafalia and Naheola Formations runs through Calhoun City. The Naheola is the easternmost formation in the North Central Hills province. About 5 miles east of Calhoun City, MS 8 crosses into the Porters Creek clay and the Flatwoods physiographic province. The Flatwoods province here is approximately 12.5 miles wide. About 1 mile east of the MS 15 junction, MS 8 crosses the southern tip of the Pontotoc Ridge province. The city of Houston is located on the Cretaceous-Tertiary boundary. The

A 35-foot-tall, east-facing exposure of the finer-grained—primarily silt—portion of the Tallahatta Formation capped by approximately 4 feet of loess. The outcrop, located on the north side of MS 8 about 200 yards east of the junction with I-55, also features 10 feet of ironstone below the loess.

Finely laminated paper shale at the sand pit at the junction of MS 8 and MS 333. The abundant flakes of muscovite mica give the shale a glittery appearance.

Paleocene-age (Tertiary) Clayton Formation rests on top of the Cretaceous-age Owl Creek sand and Prairie Bluff chalk.

In this area, the Clayton Formation is primarily sand with a few interbedded limestone beds. Houston, along with Pontotoc, New Albany, Ripley, Walnut, and Starkville, was likely settled because of springs that flowed from sand at the base of the Clayton. The Prairie Bluff is silt- and clay-rich chalk deposited in and along the margins of a sea that covered much of southern Mississippi at the end of Cretaceous time. In northern Mississippi the Ripley Formation chalk transitions to nearshore sand of the Coon Creek Tongue and the McNairy and Chiwapa Members. The change from more easily weathered carbonate to more resistant sand is the reason the Pontotoc Ridge physiographic province is where it is.

US 82
STARKVILLE—COLUMBUS—ALABAMA
36 miles

Between Starkville and the Tennessee-Tombigbee Waterway, just west of Columbus, US 82 travels on the Selma Group, which underlies the Black Prairie physiographic province. Sand and chalk formations, deposited in a shallow sea that completely inundated the Mississippi Embayment, make up this group. Note the gently rolling prairie terrain and the predominance of eastern red cedar. The underlying chalk weathers to readily form rolling hills. Unlike sand composed of silica, the calcite-rich chalk is chemically broken down by rainwater, leaving nothing solid to transport away. The cedar trees tolerate the alkaline soils that form from the weathered chalk and outcompete the acid-soil-loving pines that grow in much of Mississippi. Hills created by more resistant Ripley Formation sand briefly interrupt the gently rolling terrain between the MS 25 and MS 182 junctions at Starkville.

Mississippi State University

From US 82 in Starkville, MS 12 takes you to Mississippi State University. There is a 25-foot-tall exposure of Ripley Formation sand on the northwest side of MS 12, beside and behind a hotel, about 1 mile from the US 82 junction. This sand was deposited in coastal waters during Late Cretaceous time. The skull, lower jaw, and teeth of a Cretaceous crocodile (*Eothoracosaurus neocesariensis*) were discovered in a similar Ripley Formation outcrop in Oktibbeha County in 1973. This specimen remains the most complete fossil of the species found in North America.

Mississippi State University's Department of Geosciences houses the Dunn-Seiler Museum in Hilbun Hall. Established in 1946 to preserve and care for the extensive

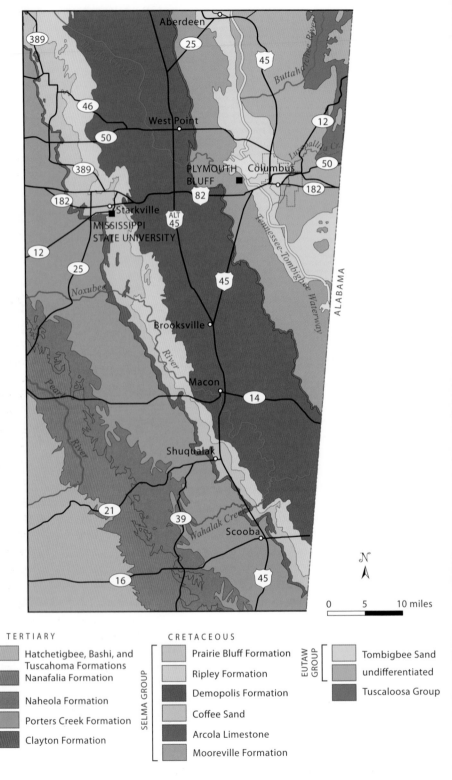

Geology along US 82 between Starkville and the Alabama state line.

geologic collections made by faculty and staff, it is the only museum in Mississippi dedicated solely to geology. Through the years, the collections have grown to hold approximately fifty thousand vertebrate, invertebrate, and plant fossils from around the world.

Instructors use an exposure of Prairie Bluff Formation chalk along Barr Avenue on campus as an outdoor laboratory for students. The chalk is nearly 60 feet thick here and represents the last widespread deposition of chalk in North America. It was deposited in a Cretaceous-age sea that covered Mississippi and much of the central plains. The chalk, which crops out across campus, is rich with fossil clams, snails, and many other shallow-water organisms.

Plymouth Bluff

The Plymouth Bluff Environmental Center, situated on the west bank of the Tennessee-Tombigbee Waterway, features a conference facility, picnic areas, hiking trails, and a museum featuring the geologic and cultural history of the area. Because of its historic

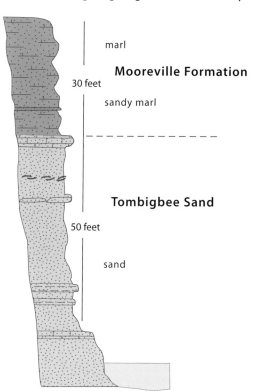

and geologic significance, the site was preserved when the Tennessee-Tombigbee Waterway was constructed. It was listed on the National Register of Historic Places in 1980. The bluff itself is composed of 50 feet of Tombigbee Sand of the Eutaw Group beneath 30 feet of the Mooreville Formation. The sand of the Tombigbee, deposited offshore in shallow water, gradually gave way to deeper-water marl of the Mooreville as sea level rose during Cretaceous time. The units contain an abundance of fossils, including mollusk shells and shark teeth, thus the site is an excellent laboratory for geology and paleontology students.

The geology of Plymouth Bluff. (Modified after Russell 1986.)

Between the MS 12 junction and the eastern overpass at MS 182, at Mississippi State University, US 82 travels on Quaternary-age floodplain and terrace alluvium of Sand Creek.

From the Tennessee-Tombigbee Waterway to the Alabama state line, US 82 travels through Eutaw Group sand. However, Quaternary-age floodplain and terrace alluvium of the waterway (Tombigbee River) and its tributaries interrupt the sand. A large area of alluvium stretches from the waterway to about 2 miles west of the Alabama state line. The Tombigbee River deposited it as it migrated across the land creating the almost flat topography. Sand and gravel mines are often visible on the north side of US 82, just west of the MS 50 junction, taking advantage of the river's work.

The contact between modern, gravelly alluvium and underlying Eutaw Group sand can be seen in the channel of Luxapallila Creek, a tributary of the Tombigbee River, from the north side of US 82, about 1 mile east of the junction with Military Road.

US 45
MERIDIAN—BROOKSVILLE—
COLUMBUS—ABERDEEN
114 miles
See the map on page 84.

From Meridian to Aberdeen, US 45 travels from south to north through the North Central Hills, Flatwoods, and Black Prairie physiographic provinces. The highway crosses the floodplain and terrace alluvium of several notable streams. The largest of these deposits belongs to the Tombigbee River, at and to the north of Columbus.

From I-20 to just south of Shuqualak and the junction with MS 39, US 45 traverses sands of the Paleocene-Eocene Wilcox Group and then sands and clays of the Paleocene-age Midway Group. The lower Naheola Formation and Porters Creek Formation of the Midway Group have more clay than the Wilcox Group formations, resulting in a decrease in relief due to higher rates of erosion. This change is visible along the highway where it transitions to the more gently rolling terrain of the Flatwoods physiographic province, which is underlain by the Porters Creek Formation. The Porters Creek is mostly clay, deposited off the coast in deeper water as sea level rose in early Paleocene time. There are no notable exposures of any of the formations along this portion of US 45.

The contact between the base of the Clayton Formation and the underlying chalk of the Prairie Bluff and Owl Creek Formations is thin sand. This sandy contact marks the Cretaceous-Tertiary boundary. There is a similar sand layer in Texas, and research indicates that a tsunami, generated when the Chicxulub meteorite hit the Earth just off the coast of what is the Yucatán Peninsula today, deposited it. Recent research, however, indicates that the sand in Mississippi is not old enough to be correlated with the tsunami. This meteorite impact was also likely responsible for triggering the mass extinction event that happened at the end of Cretaceous time, including the demise of the dinosaurs. Water flowing from springs in the basal Clayton sand may be why settlers chose the locations of Shuqualak, Starkville, Houston, Pontotoc, New Albany, Ripley, and Walnut for settlement.

From just south of Shuqualak to the Tennessee-Tombigbee Waterway at Columbus, US 45 travels on the Cretaceous-age Selma Group, which is made up of chalks of the Prairie Bluff, Ripley, Demopolis, and Mooreville Formations. Collectively, these chalks make up the Black Prairie physiographic province. Note the gently rolling prairie terrain, related to the ease with which chalk weathers (dissolves), and an increase in native junipers and cedar trees, which thrive in the alkaline soils generated by the chalks. The chalk formations in this area are almost indistinguishable from one another.

The Black Prairie was named after the black, fertile soils that form from the underlying chalks. The chalk is composed primarily of calcium carbonate. From 100 to 66 million years ago, during Cretaceous time, warmer global temperatures and higher sea level resulted in worldwide chalk deposition. Planktonic calcareous algae, known as *coccolithophorids*, flourished during this period. The single-celled algae secreted calcareous plates called *coccoliths* for protection. When the microorganisms died, the plates fell to the shallow seafloor, accumulated, and eventually were preserved as chalk. One of the world's most notable exposures of chalk is Seven Sisters, on the southern coast of England. The Cretaceous chalks extend beneath the English Channel and compose the rock unit in which the 31-mile Channel Tunnel, or Chunnel, that connects France and England was built.

Geologists have interpreted some of the steep slopes between Scooba and Macon as fault scarps, nearly vertical faces that develop when the rock on one side of a fault drops down or rises relative to the rock on the other side. Steep slopes alone are not sufficient evidence to locate a fault, since the slopes could

An outcrop of Demopolis Formation chalk just west of US 45 on MS 14. The outcrop features abundant vertical fractures, oyster fossils, and pyrite concretions (inset). The vertical fractures formed in response to widespread uplift of the area. Pyrite nodules form when iron dissolved in oxygen-depleted groundwater combines with sulfur and forms iron sulfide.

also be the result of erosion. The rock units must be displaced. The north-facing valley walls along the Noxubee River, Shuqualak Creek, and Wahalak Creek are all suspected to be on the up-thrown side of normal faults, meaning that rocks of a similar age are higher than they are on the other side of the fault. While mapping the south slope of Wahalak Creek, halfway between Scooba and Shuqualak, geologists noted 50 feet of displacement in the Porters Creek and Clayton Formations.

West of Columbus US 45 joins US 82 for about 4 miles. See the US 82: Starkville—Columbus—Alabama road log for a description of this portion of US 45.

North of the US 82 junction in Columbus, US 45 travels on alluvium of the Tombigbee River for about 2 miles before traveling through isolated, hilly

Kemper County Energy Facility

Mississippi Power is taking the combustion of lignite to a whole new level in Kemper County. Their facility, the first of its kind in the world, features a lignite mine and a 582-megawatt, state-of-the-art integrated gasification combined cycle (IGCC) electric power plant. At the plant, lignite is subjected to elevated temperature and pressure without actually combusting. This produces synthesis gas, the fuel that powers the turbines that generate electricity; its combustible components are hydrogen and methane. During the gasification process, 65 percent of the carbon dioxide emissions are captured, making the power plant's emissions equivalent to a traditional natural gas plant of the same size. The captured carbon dioxide can be used to pressurize mature oil fields to increase recovery. Some of the other waste products from the gasification process will be converted into useful chemicals and sold for industrial use. In addition to recovering contaminants from the waste stream, the facility uses wastewater from Meridian, 30 miles to the southeast, as its cooling water. Although the wastewater must be treated before use, using it eliminates reliance on precious groundwater. The mine and power plant are open for tours with advance notice.

The 8.5-foot Tuscahoma "J seam" at the Kemper County mine is the first dark layer and is separated by 3 feet of silty clay from the 1-foot-thick "J1 seam." The inset features the sharp contact between the J seam and crossbedded Tuscahoma sand. The J seam can be buried by 20 to 120 feet of sand and clay (overburden) that must be removed before it can be mined. —Courtesy of North American Coal

Looking south from the southbound lane of US 45 about 4 miles south of the junction with MS 14 (about 1 mile north of the MS 145 junction north of Shuqualak). Based on the position of a fault determined from well records, the slope could be the remnants of an eroded fault scarp (dashed line).

terrain of Eutaw Group sand. The Eutaw outcrop is only a few square miles in size. Stream piracy likely isolated it. This geologic phenomenon is called "piracy" because a river "captures" water from a different watershed as the river expands its own watershed. At one time the Tombigbee River flowed east of Columbus, but a former river captured its flow and rerouted it to west of Columbus, its current position. It's possible that the rerouting occurred during a major flood event, but regardless the channel shift spared the Eutaw hills outcrop from erosion and isolated it among alluvial deposits in all directions. The Eutaw Group sands were deposited in delta, estuary, and shelf environments during Cretaceous time.

In exposures along Luxapallila Creek, a lag deposit of the Tombigbee Sand of the Eutaw Group yields fossils of sharks, turtles, raptors, and other vertebrates. Lag deposits comprise the relatively larger sediments that are left behind after wind or water currents have removed finer particles. The Tombigbee Sand has produced more dinosaur fossils than any other unit in the state.

From about 5 miles north of the US 82 junction, US 45 travels along a terrace of the Buttahatchee River before dropping off into the modern floodplain, about 11 miles north of US 82. US 45 crosses terraces and floodplains of the Tombigbee River and its tributaries all the way to the Tennessee-Tombigbee Waterway (Tombigbee River), at Aberdeen. The lack of igneous and metamorphic rocks in the state severely limits its industrial mineral industry. However, Mississippi's climate enhances weathering, and numerous rivers have eroded, transported, and deposited huge volumes of sand and gravel through time, making it the state's leading industrial mineral. There are numerous sand and gravel mining operations in the area that extract deposits of both the Buttahatchee and Tombigbee Rivers.

From the Tennessee-Tombigbee Waterway to the MS 8W junction at Aberdeen, US 45 travels through about 3 miles of Eutaw Group sand and Quaternary-age alluvium. When crossing the waterway in Aberdeen, the hill in the distance to the northwest is a bluff of Eutaw Group sand locally referred to as Blue Bluff. The Eutaw contains glauconite, which, when unweathered, provides a greenish to bluish color. When weathered, the iron in glauconite oxidizes and turns bright red. Early settlers likely named the hill after digging holes in it and encountering the unweathered sand. Geologists cannot always judge an outcrop by its appearance. Some rock units are very distinctive, but most of the time the outer surface of an outcrop looks very different than what lies just below the surface. That's why a geologist's favorite tool is the rock hammer. Upon encountering an outcrop, the first thing a geologist does is break the rock to see a fresh surface. Because glauconite forms on continental shelves only, it is a good indicator of the environment it formed in.

US 45 ALTERNATE
Brooksville—West Point—Shannon
65 miles
See the map on page 84.

The entire stretch of US 45 ALT between Brooksville and Shannon travels on Cretaceous-age Demopolis Formation chalk of the Selma Group, which is part of the Black Prairie physiographic province. However, in the vicinity of MS 50, at West Point, US 45 is on Quaternary-age floodplain and terrace alluvium of Tibbee Creek. The downtown portion of West Point east of US 45 ALT is on another, higher gravel terrace of Tibbee Creek.

The Black Prairie is named for the black fertile soils that form from the underlying chalks, which are composed primarily of calcium carbonate. From 100 to 66 million years ago, during Cretaceous time, warmer global temperatures and higher sea level resulted in worldwide chalk deposition. Planktonic calcareous algae known as coccolithophorids flourished during this period. The single-celled algae secreted calcareous plates called coccoliths for protection. When the microorganisms died, the plates fell to the shallow seafloor, accumulated, and eventually were preserved as chalk.

Due to its thickness and the ease with which it is excavated, the chalk deposits of eastern Mississippi received considerable attention from developers in the late-twentieth century. In 1987 Mississippi unsuccessfully promoted its chalk as a suitable foundation for the Superconducting Super Collider. However, a site in Texas, in the Austin Chalk, was approved, but construction halted after fewer than 15 miles were completed. In 1991 Brooksville and Shuqualak were promoted as good locations for two hazardous-waste landfills, but neither was constructed.

North of West Point, US 45 ALT crosses into Monroe County. This region is the heart of the Black Warrior Basin, one of the two prominent oil and gas provinces in the state. Several natural gas wells and a gas plant are visible to the east about 3 miles north of the MS 25 junction. Natural gas was first discovered in Mississippi at Amory Field in Monroe County. The first production was in the Mississippian-age Carter Sandstone of the Floyd Shale.

An economic accumulation of hydrocarbons requires a source, a reservoir, and a seal, or trap. The Carter Sandstone was likely deposited as a marine bar just off the coast. Marine bars make good reservoir rock because wave activity inhibits clay from being deposited with the sand grains. As a result, the open spaces between the grains, called pores, are often preserved and connected. This makes the rock permeable. If hydrocarbons are generated in a nearby source rock (one rich in organic matter) and pass through the permeable rock, the rock can become a reservoir. Because hydrocarbons float on water, they seek the highest position possible when migrating through porous rock. If a trap has formed (by faulting, folding, or other processes), the hydrocarbons stop migrating, and the reservoir can fill with gas and oil over time.

The Amory Field was exhausted of natural gas by 1938, but the same properties that made it a good natural reservoir also make it a good place to store natural gas. As long as the pressure of the stored gas does not rupture the reservoir's trap, gas can be added or removed as needed. The Amory Field, once used for gas production, is now used to store natural gas at the Monroe Gas Storage Facility, in Amory.

US 45
ABERDEEN—TENNESSEE
91 miles

From the MS 8W junction at Aberdeen, US 45 travels through Tombigbee Sand of the Eutaw Group for 4 miles. It then passes through Selma Group chalks of Cretaceous age, which are part of the gently rolling terrain of the Black Prairie physiographic province. East of Aberdeen the Eutaw Group includes an 8-foot-thick seam of bentonite that is mined for cat litter. Bentonite is clay that forms from volcanic ash. Both the Jackson and Midnight volcanoes were active in the Mississippi Embayment when the ash was deposited, so it is difficult to say which volcano was responsible for the ash.

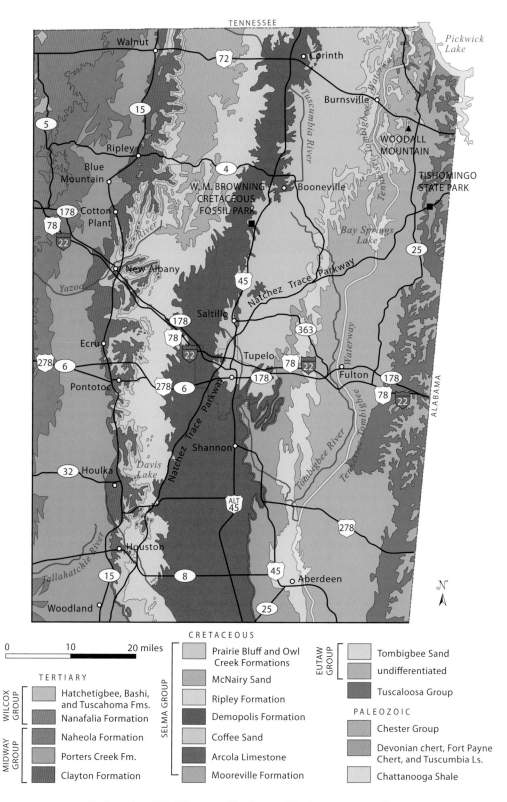

CRETACEOUS

TERTIARY

WILCOX GROUP
Hatchetigbee, Bashi, and Tuscahoma Fms.
Nanafalia Formation

MIDWAY GROUP
Naheola Formation
Porters Creek Fm.
Clayton Formation

SELMA GROUP
Prairie Bluff and Owl Creek Formations
McNairy Sand
Ripley Formation
Demopolis Formation
Coffee Sand
Arcola Limestone
Mooreville Formation

EUTAW GROUP
Tombigbee Sand
undifferentiated
Tuscaloosa Group

PALEOZOIC
Chester Group
Devonian chert, Fort Payne Chert, and Tuscumbia Ls.
Chattanooga Shale

0   10   20 miles

Geology along US 45 between Aberdeen and the Tennessee state line.

Through Tupelo, US 45 travels on the alluvial deposits of Town Creek. The original settlement of Tupelo, visible to the west, is on the Coffee Sand, deposited as deltas advanced to the southwest over Mooreville Formation chalk during Cretaceous time. In this area the Arcola Limestone, a thin limestone unit, is differentiated from the underlying Mooreville Formation chalk, but it is not exposed along US 45.

Once off the floodplain of Town Creek, US 45 becomes hilly as it traverses the Coffee Sand to about 1 mile north of the MS 145 junction at Saltillo. The hills of the Coffee Sand, and of the Eutaw Group to the east, are part of the

W. M. Browning Cretaceous Fossil Park

W. M. Browning Cretaceous Fossil Park is one of the few parks dedicated to a geological theme in the state. About 1 mile north of the MS 30W junction at Frankstown, exit east onto the unmarked county road (7450). Once off the exit ramp, the entrance to the park is immediately to the south.

There is an exposure of Coffee Sand in the wall of Twentymile Creek at the park. The boulders, or concretions, in the wall and along the creek were not rounded by stream transport; they were created through the process of selective cementation within the Coffee Sand. Why concretions were cemented in this way is not understood.

The main attraction at the park is the fossil-rich sand, informally named the "Frankstown Sand," in the lowest portion of the overlying Demopolis Formation

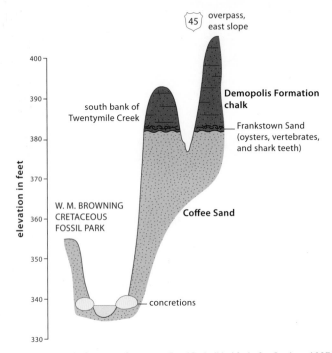

The geology at W. M. Browning Cretaceous Fossil Park. (Modified after Dockery 1997.)

chalk. About 75 million years ago sea level rose and inundated the Coffee Sand. The wave action winnowed away the lighter sand particles, leaving behind copious fossils that it wasn't powerful enough to remove. The Frankstown Sand is what is called a lag deposit. It's about 1 foot thick and lies on top of the Coffee Sand and beneath an oyster bed of the Demopolis Formation chalk. The fossils of shark teeth, bony fish, oyster shells, and more than sixty species of vertebrates have been documented in this layer. Reportedly, crocodile and hadrosaur (duck-billed dinosaurs) fossils have also been found. As the shoreline progressed to the northeast and the water grew deeper, sand, silt, and clay gave way to chalk—the remains of planktonic calcareous algae—which was deposited on the shelf of a warm tropical sea.

A 30-foot-tall exposure of Coffee Sand in the south wall of the creek bed at W. M. Browning Cretaceous Fossil Park. Concretions in the creek bed range up to 5 feet in diameter. The Frankstown Sand is not exposed here, but fossils weather and erode from the sand and come to rest in the stream.

Tombigbee and Tennessee River Hills physiographic province. This province has more relief because its formations are resistant to erosion.

For roughly the next 17 miles, US 45 travels along the contact of the Demopolis Formation chalk and the Coffee Sand, both of the Selma Group. The sections of road with more relief are on the Coffee, whereas the sections of gently rolling prairie are on the Demopolis, which weathers relatively easily.

From Booneville to the Tennessee state line, US 45 travels primarily on the Demopolis Formation chalk, once again back on the Black Prairie physiographic province. However, from the Tuscumbia River crossing north to the US 72 junction, US 45 is on alluvial deposits of the Tuscumbia.

US 72

Tennessee—Corinth—Alabama

89 miles
See the map on page 93.

US 72 passes through several physiographic provinces in northeast Mississippi. Although each represents different geology, it is sometimes difficult to distinguish where one changes to the other.

From the Tennessee state line to 10 miles east of the junction with MS 5 (34 miles total), US 72 travels primarily in the North Central Hills physiographic province on sand, lignite, and clay of the Paleocene-Eocene-age Wilcox Group. Though lignite prospects were identified in Benton County, neither has been exploited. About 11.5 miles east of MS 5, just over the Benton-Tippah county line, US 72 crosses from the Naheola Formation, composed of sand deposited on an alluvial plain and in an estuary, to the marine clay of the Porters Creek Formation. The Flatwoods physiographic province has developed on the Porters Creek, which erodes relatively easily. The bottom of the Porters Creek is a limestone rich in marine gastropod (snail) fossils. The distinctive shells of *Turritella mortoni* taper to a point and are tightly coiled. As sea level rose, the limestone sediment was replaced by fine clays deposited well offshore in deeper water. Unfortunately there are no good outcrops still exposed that afford fossil collecting.

Much of US 72 between MS 15 and US 45 travels on the Cretaceous-age Selma Group, first on the Owl Creek Formation and then the Ripley Formation. The sandy units of the Ripley Formation (Chiwapa and McNairy Members and the Coon Creek Tongue) are the erosion-resistant units that make up the Pontotoc Ridge physiographic province. The Coon Creek Tongue is fossil-rich in Mississippi but best known for an area in adjacent McNairy County, Tennessee, which is the home of the Coon Creek Science Center. The science center showcases many marine and vertebrate fossils found in the Coon Creek Tongue.

From 1 mile east of the County Road 604 junction (to Kossuth) to about 3.5 miles east of US 45, US 72 travels on Demopolis Formation chalk, also of the Selma Group. This chalk makes up the Black Prairie physiographic province. Composed of the remains of microscopic calcareous plankton, the chalk was deposited in a relatively warm sea during Cretaceous time. Just west of its junction with US 45, US 72 is on the almost flat Quaternary-age terrace and floodplain alluvium of the Tuscumbia River, which differs from the gently rolling terrain of the Demopolis.

East of the Tuscumbia River to the Alabama state line there is a notable increase in relief as US 72 travels through sand and gravel deposits of the Tombigbee and Tennessee River Hills physiographic province. From west to east, the highway travels through the Coffee Sand, Eutaw Group, and Tuscaloosa Group, all of Cretaceous age. The Tuscaloosa sand and gravel were deposited in rivers that

A 35-foot-tall cliff of the Ripley Formation (mostly sand) on the north side of US 72 about 5 miles east of the Tippah-Alcorn county line. The Ripley was deposited on a shallow marine shelf landward of the chalk deposition that was taking place in central Mississippi.

An impressive 80-foot-tall exposure of Tuscaloosa Group sand and gravel less than 1 mile west of the Alabama state line, in the northeast corner of the junction of US 72 and County Road 172.

drained into the Mississippi Embayment from exposed Paleozoic-age bedrock in Tennessee. Both the Eutaw and Coffee contain glauconite, a mineral that typically forms in marine sediment and is enriched in iron, potassium, and magnesium. When exposed in outcrop it weathers to the red to reddish-brown color seen in most exposures of these rocks.

Woodall Mountain

At 807 feet, Woodall Mountain, east of Burnsville and south of US 72, is the highest point in Mississippi. Mississippi is not the state with the lowest average elevation; it is actually the fourth lowest! The total relief relative to surrounding terrain is 396 feet. The top of Woodall Mountain is composed of a sandstone remnant of the resistant Coffee Sand, deposited as a delta advanced to the southwest during Cretaceous time. Woodall Mountain can be accessed via MS 25.

INTERSTATE 22/US 78
New Albany—Tupelo—Alabama
49 miles
See the map on page 93.

New Albany has been developed on the Paleocene-age Clayton Formation and the Cretaceous-age Owl Creek and Ripley Formations. All three formations are sandy and very similar to each other in appearance. This region, part of the Pontotoc Ridge physiographic province, is hilly and has more relief because the sandy formations are resistant to erosion.

The Clayton sand was deposited in shallow water during the first marine transgression into the Mississippi Embayment. It is the oldest Paleocene-age rock unit exposed in Mississippi and often contains fossils of Cretaceous age that were worked loose from the underlying Owl Creek sands and Prairie Bluff chalk as the sea rose across the eroded Cretaceous landscape.

During the Late Cretaceous the Owl Creek and Prairie Bluff were deposited at the same time but in different environments. North of New Albany, the Owl Creek Formation is sandy and was closer to a source of runoff from the coast. In New Albany, and to the south, the bedrock is the Prairie Bluff, a chalk that was deposited in water farther from the influence of runoff.

The Ripley Formation is composed of fossil-rich sands deposited in shallow coastal waters. Its most famous member is the Coon Creek Tongue. The Coon Creek is the bottom, sandy layer that contains abundant mollusk and crab fossils. Excavation activity in the New Albany area often garners interest because fossils of the Coon Creek might be unearthed.

Between exit 64, at the MS 15 junction, and mile marker 69, I-22/US 78 crosses the Prairie Bluff and Clayton Formations several times. The units are very thin, with the younger Clayton lying on top of the Prairie Bluff on hilltops. I-22/US 78 then transitions onto chalk of the Demopolis Formation, which it travels over to about 2.5 miles east of the MS 9 junction. The chalk underlies the portion of the Black Prairie physiographic province that lies west of Tupelo; it is composed of the remains of calcareous algae that lived in a relatively warm sea. The chalk is fossil rich in many places; oyster shells are abundant in it. Note the gently rolling prairie terrain and the abundance of native juniper and eastern red cedar trees; the topography has developed because the chalk weathers and erodes easily, and the trees thrive in the alkaline soil it creates. There are several exposures on both sides of I-22/US 78 near mile markers 80 and 81.

Between mile markers 84 and 94, I-22/US 78 travels through the Coffee Sand of the Selma Group; the sand comprises the Tombigbee and Tennessee Hills physiographic province at Tupelo. Note the increased relief, which is the result of the sand's resistant nature. The almost flat terrain of Quaternary-age floodplain and terrace alluvium of Town Creek and one of its tributaries interrupts the hilly terrain from just west of the US 45 junction to exit 87.

Between mile markers 94 and 97, I-22/US 78 travels on Mooreville Formation chalk, which underlies the portion of the Black Prairie that lies east of Tupelo. Note again the gently rolling prairie and the abundance of native junipers and eastern red cedars.

A 25-foot-tall exposure of Demopolis Formation chalk on the south side of I-22/US 78 at exit 81 (Tupelo).

Between mile marker 97 and the Alabama state line, I-22/US 78 travels through Eutaw Group sand and Tuscaloosa Group sand and gravel, both deposited by rivers in shoreline environments during Cretaceous time. Note the increase in relief, the result of the formations' resistance to erosion. These formations make up the Tombigbee and Tennessee River Hills physiographic province. Exit 97 is located in the Tombigbee Sand of the Eutaw Group. The Tennessee-Tombigbee Waterway and the Quaternary-age floodplain and terrace alluvium of the Tombigbee River interrupt the hilly terrain at mile marker 105. There are small exposures of Tuscaloosa Group gravel at the rest area just east of mile marker 115, on the north side of I-22/US 78.

US 278/MISSISSIPPI 6
Pontotoc—Tupelo
42 miles
See the map on page 93.

From the MS 15 junction, just northwest of Pontotoc, to 1 mile east of the MS 9 junction, US 278/MS 6 passes through sand of the Paleocene-age Clayton Formation and silty chalk of the Cretaceous-age Prairie Bluff Formation. The Prairie Bluff here is approximately 70 feet thick, and to the north it grades into the Owl Creek Formation, in Union County. The road then traverses the Ripley Formation of the Selma Group to the junction of US 278/MS 6 and MS 76. These sandy units all contain glauconite, which weathers to a red to brown color; they make up the Pontotoc Ridge physiographic province.

US 278/MS 6 then transitions through chalks of the Demopolis Formation, which make up the Black Prairie physiographic province, and follows them to the Tupelo National Battlefield in Tupelo. Note the gently rolling terrain and the increase in native junipers and eastern red cedars. They thrive on the alkaline soils created by the chalk, which is made up of the remains of planktonic calcareous algae that lived in Cretaceous seas. Pine species, common in much of Mississippi, cannot tolerate alkaline soils and are outcompeted by the cedars and junipers.

Between the park and the junction with US 45, MS 6 travels through downtown Tupelo on the Coffee Sand of the Selma Group, except where it crosses the Town Creek floodplain, which consists of alluvium. For a short period of time the Coffee Sand and Mooreville Formation chalk were deposited simultaneously, the Coffee by rivers at and near the shoreline and the Mooreville offshore, away from the influence of the sand and clay. As the Coffee Sand deltas extended into the Mississippi Embayment, they overrode the Mooreville chalk. The Coffee Sand forms the western portion of the Tombigbee and Tennessee River Hills physiographic province and is the formation on top of Woodall Mountain, the highest elevation in the state, in Tishomingo County to the north.

MISSISSIPPI 15
Woodland—Tennessee
92 miles
See the map on page 93.

From Woodland to 4 miles north of the MS 8 junction at Houston, MS 15 travels on clay of the Porters Creek Formation, part of the Paleocene-age Midway Group. There is a noticeable change in relief as you approach the MS 32E junction, north of Houston, as the highway crosses onto more resistant Clayton Formation sand, deposited by the first transgressive seas of Paleocene time. The Clayton sand characteristically contains abundant Cretaceous-age fossils torn from the underlying Prairie Bluff and Owl Creek Formations.

About 2 miles north of the MS 32W junction at Houlka is the access to Davis Lake. County Road 413 is a scenic byway that travels within the Tombigbee National Forest, mostly through Clayton Formation sand and then Cretaceous-age chalks of the Selma Group at Davis Lake.

From the turnoff to Davis Lake to just south of the US 78 junction at New Albany, MS 15 travels in and out of, and along the contact between, Clayton Formation sand and Prairie Bluff chalk. The Prairie Bluff was the last of the Cretaceous chalks deposited in Mississippi and was deposited as far north as New Albany. Between the MS 346 junction and just north of the MS 345 junction at Ecru, US 78 crosses notable Quaternary-age terrace and floodplain alluvium of the Little Tallahatchie River.

From just south of the US 78 junction to north of Cotton Plant, MS 15 travels in and out of, and along the contacts between, the Clayton Formation, Owl Creek Formation, and Ripley Formation. North of New Albany, Prairie Bluff chalk transitions into Owl Creek sand. Both the chalk and the sand were deposited at the same time, but the Owl Creek was closer to shore and thus contains sand that was too dense to be transported very far from shore. The two formations are said to *intertongue*. At a shifting boundary, such as a coastline, a shift in the depositional environment can cause a "tongue" of one formation to extend into a different one. For example, imagine that chalk is being deposited on the seafloor while sand is deposited along the shoreline. If sea level drops, sand deposition moves seaward; if sea level then rises, chalk deposition moves inland. In cross section, what you end up with is a tongue-shaped deposit of sand that extends into the chalk. This term was first used to describe the geology of several units in Mississippi. Without careful subsurface investigations, complex sedimentary relationships, such as strata that intertongue, are difficult to interpret.

The Ripley Formation is composed of fossil-rich sands deposited in shallow coastal waters. Its most famous member is the Coon Creek Tongue, the bottom, sandy layer that contains abundant mollusk and crab fossils. Excavation activity in the New Albany area often garners interest because fossils of the Coon Creek might be unearthed. It is difficult to distinguish between the sandy Clayton, Owl Creek, and Ripley Formations, which form the Pontotoc Ridge physiographic province in this area.

About 2 miles north of Cotton Plant the highway returns to gently rolling terrain of the more easily eroded Porters Creek Formation clay, part of the Flatwoods physiographic province. Just north of the MS 2 junction in Blue Mountain, the high hill to the west is composed of Wilcox Group sands of Paleocene-Eocene age.

From about 2 miles north of the MS 2 junction to the Tennessee state line, MS 15 travels in and out of, and along the contacts between, the clay and sand of the Porters Creek Formation and mostly sandy marl of the Clayton Formation. In Tippah County, the Clayton transitions, from south to north, from primarily sand to primarily marl with beds of the Chalybeate Limestone Member. The type section for this limestone is about 2 miles southeast of MS 15 on MS 354, in a ravine north of the town of Chalybeate. The limestone in the Chalybeate is known for its abundant *Turritella* gastropod (snail) fossils, which are tightly coiled shells that taper to a point.

Along this stretch, MS 15 travels on the western edge of Pontotoc Ridge. Sandwiched between it and the North Central Hills, to the west, is the Flatwoods physiographic province. Although the Porters Creek Formation—a clay that erodes relatively easily—underlies the Flatwoods, providing the featureless landscape described by its name, the width of the outcrop narrows to just a few miles here. So, north of Ripley even the Flatwoods province is hilly, and the distinctions between the North Central Hills, the Flatwoods, and Pontotoc Ridge become difficult to note.

NATCHEZ TRACE PARKWAY
HOUSTON—ALABAMA
80 miles
See the map on page 93.

From the MS 8 junction near Houston to mile marker 239, the Natchez Trace Parkway traverses Cretaceous-age formations (Owl Creek sand, Prairie Bluff chalk, and Ripley Formation sand) of the Pontotoc Ridge physiographic province. The chalk was deposited offshore in a sea that engulfed the state, and the sand was deposited very near the shore in a delta. All of the formations belong to the Selma Group.

Between mile markers 239 and 262, the Trace crosses the Black Prairie physiographic province, which is underlain by the Demopolis Formation chalk. The chalk is composed primarily of calcium carbonate, which is chemically weathered by precipitation and slowly dissolves through time, creating the gently rolling prairie terrain. Native eastern red cedars tolerate the alkaline soil created from the chalk as it weathers, which correlates with their abundance in the Black Prairie province. From 100 to 66 million years ago, during Cretaceous time, warmer global temperatures and higher sea level resulted in worldwide chalk deposition. Planktonic calcareous algae known as coccolithophorids flourished during this period. The single-celled algae secreted calcareous plates called coccoliths for protection. When the microorganisms died, the plates fell to the shallow seafloor, accumulated, and eventually were preserved as chalk. Chalk outcrops are visible along the shores of Davis Lake, accessed via a scenic byway near mile marker 243. The byway courses west from the Trace through chalks of the Selma Group.

Between mile markers 262 and 302, the Trace travels through the Tombigbee and Tennessee River Hills physiographic province, composed of Quaternary-age terrace deposits; Cretaceous-age Coffee Sand, Eutaw Group sand, and Tuscaloosa Group gravel; and sandstone, shale, and limestone of the Pride Mountain Formation and the Hartselle Sandstone, both of the Mississippian-age Chester Group. The Pride Mountain was deposited on a stable, shallow shelf, and the Hartselle was either a barrier island complex or a delta. Hartselle Sandstone is exposed at the base of the Whitten Lock and Dam of the Tennessee-Tombigbee Waterway; you can see it from the bridge near mile marker 293.

Tennessee-Tombigbee Waterway

A river's natural path is the one of least resistance. Humans, however, can intervene in this path in two ways: we fight nature by keeping a river captive in its channel, or we connect two rivers to shorten the distance to the ocean. Both endeavors are expensive in terms of dollars and energy expended to maintain the imbalance created by denying rivers their natural tendencies. How we've related to the Mississippi River is an example of the first intervention. If nature had its way, the Mississippi would be

Mississippian-age Hartselle Sandstone exposed at base of the Whitten Lock and Dam.

flowing through Morgan City, Louisiana. The Tennessee-Tombigbee is an example of the latter. This waterway was cut through the oldest and best-consolidated rocks in the state. Humans excavated rocks that nature sought ways around.

French trapper Marquis de Montcalm is credited for being the first to record—in the late 1700s—interest in constructing a waterway linking the Tennessee and Tombigbee Rivers. The idea was raised frequently over the next two centuries, until the waterway received congressional approval in 1946. Construction of the economically and environmentally controversial waterway did not begin until 1968. When completed in 1984, the 234-mile waterway averaged 9 feet in depth and featured ten locks to lift vessels 341 feet, the total difference in elevation between the two rivers. The Tennessee-Tombigbee is the world's largest earthmoving project, requiring the removal of 310 million tons of rock and soil. Coal and timber are the primary commercial commodities shipped on the waterway and make up the bulk of the 8 million tons shipped annually. Initial estimates were that the waterway would

Tishomingo State Park

Between mile markers 304 and 305 is an entrance to Tishomingo State Park. There are numerous exposures of Paleozoic-age Hartselle Sandstone and sand and limestone of the Pride Mountain Formation, both of the Chester Group, within the park. The Chester Group rocks were deposited on a stable shelf and in a delta or barrier island complex. The Civilian Conservation Corps utilized the Hartselle Sandstone to build the park's facilities, in the 1930s, including the abutments for the swinging bridge across Bear Creek. The limestone that makes up the channel floor in Bear Creek is the Green Hill Member of the Pride Mountain Formation. It was deposited in a warm and shallow sea during Mississippian time.

The Outcroppings Trail offers an excellent opportunity to observe geology up close, including jointing, crossbedding, and an unconformity. *Joints* are fractures in bedrock along which there has been no movement. When rocks are uplifted from deep within the Earth due to tectonic forces, they become more brittle and break. The vertical orientation of the joints in the Hartselle is indicative of vertical uplift. *Crossbedding* is a depositional structure that forms in dunes and ripples. In cross section, crossbeds are concave, inclined layers that occur within sandstone or limestone beds.

An *unconformity* is a gap in the rock record, a time during which no sediment was deposited to become rock, or rock and sediment were eroded away. On the ground, unconformities are the surface where two formations of vastly different ages abut. The boundary between the Cretaceous-age Tuscaloosa gravel and the Mississippian-age

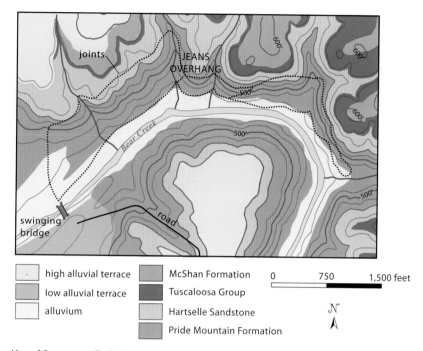

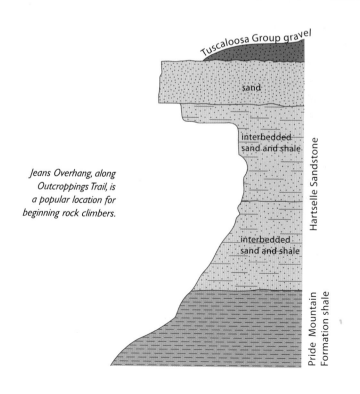

Map of Outcroppings Trail (dashed line), which is an excellent natural laboratory for geology students to observe joints, fossils, crossbeds, and contacts between the Pride Mountain, Hartselle, and Tuscaloosa units.

Jeans Overhang, along Outcroppings Trail, is a popular location for beginning rock climbers.

These linear, vertical faces developed along joints in the Hartselle Sandstone.

An approximately 10-foot-tall exposure of Hartselle Sandstone overlying Pride Mountain Formation shale between mile markers 305 and 306, just north of Bear Creek.

Hartselle Sandstone marks a great unconformity—a gap in time of just over 200 million years. In Tishomingo County the Pennsylvanian, Permian, Triassic, and Jurassic Periods are not represented by rock. In that timeframe, a mountain range the size of the Himalayas, produced by the closing of Iapetus, was removed by erosion. The Tuscaloosa gravels, deposited by streams during the final erosional stages of those Paleozoic highlands, can be seen on top of the Hartselle along Outcroppings Trail.

The Hartselle is also known for containing molds and casts of invertebrates, trace fossils, and even imprints and casts of the giant tree fern *Lepidodendron*. Molds are the hollow spaces left in rock after the remains of buried organism have decayed. Casts develop when minerals gradually fill a mold, leaving a shape that mimics the original organism. And trace fossils are impressions an organism left in soft sediment, such as burrows and tracks. The Hartselle was a favorite building stone throughout Tishomingo County and was quarried until 1982.

Cave Spring is located west of the Trace, just south of the Alabama state line, at mile marker 308. A short hike takes you to the cave and an exposure of the calcareous Southward Springs Sandstone Member of the Pride Mountain Formation. This sandstone formed in a delta environment during an episode of increased sand input to the coast. The "cave" is more appropriately a collapse feature, or sinkhole. Over time, surface water percolated through the Southward Springs Sandstone into the limestone beneath, slowly dissolving it and creating a cavern. The sandstone cap collapsed into the cavern, forming the Cave Spring feature. Caves are generally subsurface features that form as rock dissolves due to chemical weathering, not collapse.

THE DELTA:
MISSISSIPPI RIVER ALLUVIAL
PLAIN AND LOESS HILLS

Mississippi's "Delta" region should not be confused with the actual Mississippi River Delta. A delta is a landform that forms at the mouth of a river. The true Mississippi Delta is the elongated deposit of sand forming south of New Orleans and stretching into the Gulf of Mexico. The "Delta," as it is affectionately called, is actually a 7,000-square-mile plain (15 percent of Mississippi) in the western part of the state.

The Delta is formally known as the Mississippi River Alluvial Plain; its boundaries are the Mississippi River on the west, the loess bluffs on the east, Memphis to the north, and Vicksburg to the south. The alluvial plain covers all or part of seventeen counties; at its widest point it measures 70 miles.

Geologic History

The North American crust, broken and weakened by rifting in late Precambrian time, enabled the formation of the Mississippi Embayment and created an easy path to the ocean for North America's glacial snowmelt and rainwater runoff. The erosive action of the modern Mississippi and its tributaries, as well as the ancestral Mississippi and Ohio Rivers, has flattened the Delta. As recently as 85,000 years ago both the ancestral Mississippi and Ohio Rivers flowed through the Mississippi Embayment to the Gulf of Mexico. The ancestral Ohio River flowed much farther to the south than it does today, perhaps as far south as Natchez. Both rivers left behind telltale signs of their presence. Point bar deposits of sand and gravel, ridge and swale topography, abandoned courses and channels, oxbow lakes, meander scars, and splay deposits are common geomorphic features of the Delta. Although some research has been done to differentiate when and how the features were created, many of them could be from either ancestral rivers or their tributaries.

The modern configuration of the two rivers we are familiar with was established approximately 10,000 years ago. Since that time the Mississippi has been the major sculptor of the alluvial plain. The Mississippi's drainage basin now spans from Montana to Pennsylvania, and all of its water ends up flowing over the great Mississippi River Delta forming in the Gulf of Mexico, off the coast of Louisiana.

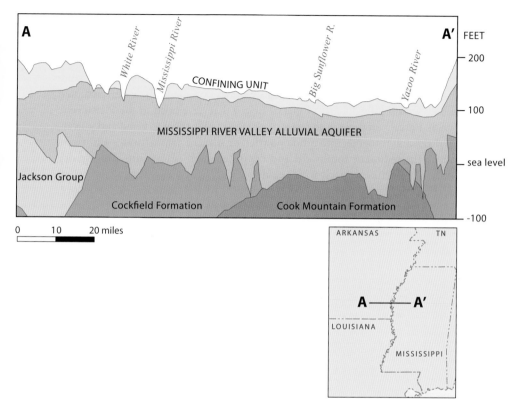

East-west cross section of the Mississippi River Valley Alluvial Aquifer.
(Modified after Renken 1998.)

When sea level dropped 410 feet during the Wisconsin glacial stage (85,000 to 11,000 years ago), the ancestral Mississippi and Ohio Rivers responded by eroding their channels deeper into Tertiary-age bedrock. As sea level rose, the scoured river valleys gradually filled back up with sand and gravel delivered by the Mississippi River. The sand and gravel deposits in that deep valley form the Mississippi River Valley Alluvial Aquifer.

The aquifer is composed of two units: an upper confining layer and the aquifer itself. The confining layer is composed of sand, silt, and clay, and the aquifer is composed of coarse gravel at its base and progressively finer sediment leading to sand at its top. The total thickness of both units ranges from 75 to 200 feet, and throughout much of the alluvial plain it ranges between 120 and 160 feet in thickness. The aquifer unit averages 110 feet in thickness and is thinnest adjacent to the loess bluffs, to the east. The overlying confining layer averages 25 feet in thickness. Approximately 98 percent of the water in the aquifer is used in agriculture to produce the majority of the state's rice, cotton, corn, soybeans, and catfish. Vicksburg is the only municipality taking advantage of the aquifer for drinking water.

LOESS

Loess was first described in the Rhine Valley in Germany in the 1830s. Deposits were first described in the Lower Mississippi Valley in 1846. There has been considerable debate about the origin of loess over the years, and although some localized deposits may have been redistributed by water, it is widely accepted to be a windblown deposit of glacial origin that developed in the Pleistocene Epoch. Entrapped rocks at the base of glaciers enable them to grind and pulverize rocks and minerals into a fine powder, or silt. Glacial meltwater and rivers moved the sediment southward and deposited it on river plains. During dry periods, prevailing westerly winds easily picked up the fine sediment, and when the wind encountered the east valley wall of the Mississippi River and its associated tributaries, it slowed down. The silt dropped out and blanketed the terrain, creating vast loess deposits.

The deposits in Mississippi, which comprise the Loess Hills physiographic province, are generally thicker on ridge crests than in the valleys, with the thickness decreasing eastward across the state. The province butts up against the Delta's eastern margin, forming a line of bluffs composed of the tan sediment. Loess erodes into almost vertical walls, thus there is significant relief where there are thick deposits, typically at the province's western edge, closest to the floodplains of the Mississippi River and its tributaries. Loess along US 61 near Vicksburg exceeds 100 feet and is among the thickest sections recorded in the United States. Coverage is spotty eastward from the bluff line, and generally only a small percentage of the area indicated on maps as being loess covered still contains loess. Fossils in a US 61 outcrop provided the oldest and youngest ages for Mississippi's loess: around 25,300 years, determined from gastropods, and 17,850 years, determined from fossil shells.

Mississippi loess is composed of 65 percent silt-sized quartz, 20 percent carbonate, 6 percent feldspar, 7 percent clay, and 2 percent other minerals. The carbonate-enriched portion features root tubules. These formed from carbonate minerals that precipitated around roots; when the roots died and rotted, mineral tubes were left. Calcite concretions form in loess as carbonate dissolved in upper horizons is precipitated in lower zones. There are several German names for the globular concretion forms, such as *losskindchen*, *lossmanchen*, and *losspuppen*. Locally they are known as "loess dolls." Concretions also form where roots have deteriorated, allowing the calcite to precipitate.

One of the most interesting characteristics of loess is its ability to maintain vertical faces. Both the upper weathered zone and the lower carbonate-enriched zone exhibit this tendency, but the carbonate interval has greater integrity. The grains within the loess are not cemented the way that grains are in many sedimentary rocks—that is, with a mineral cement. Rather, geologists attribute loess's ability to hold together to a thin film of water surrounding its sediment grains. The mechanical compaction and sloping of a loess surface greatly increases the tendency for slumping and erosion. Loess slopes that aren't cut into vertical cliffs must be sodded immediately to prevent failure. Mississippi loess is relatively permeable. Geologists believe the uniformity of

the silt-sized particles, low clay content, and presence of vertically oriented calcite tubules enable water to pass efficiently through the loess. This permeability
results in the upper weathered zone that averages 14 feet thick.

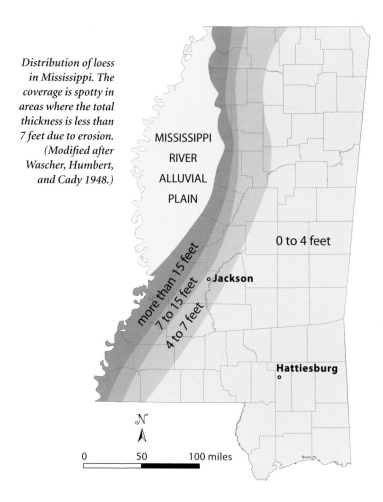

*Distribution of loess
in Mississippi. The
coverage is spotty in
areas where the total
thickness is less than
7 feet due to erosion.
(Modified after
Wascher, Humbert,
and Cady 1948.)*

The loess caps Pleistocene-age alluvial sand and gravel, as well as older sand,
clay, and limestone of Pleistocene and Tertiary age. In areas east of the bluff line,
the sand and gravel beneath the loess are described as *pre-loess terrace deposits.*
These deposits are older than the loess but younger than the Tertiary-age bedrock below. State geologists are currently remapping the entire belt of loess to
determine the extent of the pre-loess terrace deposits. It is not known which
river, or rivers, is responsible for the terraces. The likely choices are the ancestral
Mississippi and Ohio—perhaps even the Tennessee—Rivers. Sand deposits can
be dated by a relatively new technique known as optically stimulated lumines-
cence, but there has been no dedicated effort to date the terrace deposits.

RIVER DYNAMICS AND HISTORY

The difference in elevation between any two points along a channel causes water to flow. Geologists refer to the change in elevation divided by the total channel distance between two points as *gradient*. Near a river's headwaters, or where it begins, the gradient is large; near the river's mouth, where the river dumps into a larger river or the sea, the gradient is low. The gradient between the two points decreases constantly between the headwaters and the mouth. Numerous factors control gradient, such as the length of the stream, size of the basin, amount of precipitation the basin receives, type of rock the stream flows over, and underlying rock structure. Generally, high-gradient streams have straight and relatively narrow channels with rapids and waterfalls and no floodplain. Low-gradient streams develop large meanders, form very wide floodplains, and have multiple distributary channels, which branch from a main river channel and may or may not rejoin the main channel.

The average gradient of the Mississippi River between Memphis and the Mississippi-Louisiana state line is only 4 inches per mile. To put this in perspective, the average gradient of the Mississippi River through all of Mississippi is less than the amount of tilt observed when you place a single sheet of paper under one edge of a previously level 4-foot-wide surface.

As a river wears down the landscape, the stream gradient is lessened, and the relative flatness of the topography causes the river to wind widely back and forth, or meander, across its floodplain. Meanders become more pronounced as the river erodes its banks. Because water flows faster along the outer bank of the channel in a meander, this is where erosion occurs, creating what is called a *cut bank*. On the inside of a meander water flows much slower, so the river deposits the material it's transporting in a *point bar*. As the river migrates back and forth across the landscape, it continues to erode and redeposit material across its wide, flat floodplain.

The velocity and volume of water in the channel and the availability of material to erode determine how much sediment—and the maximum size of the particles—a river can carry; this is called its *load*. The load is composed of three parts: the bed load, which rolls and bounces along the bottom of the channel; the suspended load, which is carried in the flow; and the dissolved load, which cannot be seen with the naked eye. So, depending on the volume and velocity of the water and the availability of material, channel and point bar deposits can be composed of silt, sand, gravel, or a combination of all three. Mississippi rivers are brown because of suspended clay and silt, the smallest sedimentary particles.

Most of the time a river or stream transports its load to a larger river or, eventually, the ocean; the Mississippi, of course, ends in the Gulf of Mexico. However, when a river leaves its channel during flood events, all that silt and clay is deposited across the floodplain. Repeated flooding leads to the formation of elevated banks, or natural levees, along the river channel. Levees are coarse grained compared to the other overbank deposits because sand is deposited first as the river leaves its channel, then the silt and clay.

Rivers can also leave their channels through breaches in the natural levee, permitting water to flow out onto the floodplain. This process is known as *avulsion*, and the streams are called *crevasse streams*. These streams are often temporary, but they can cause the river to abandon its main channel if the gradient is low enough in a nearby basin—a type of piracy. They can also become distributaries if they occur relatively close to the river's mouth. As water leaves the confines of the river channel it loses some of its energy, so it can no longer carry coarse sediment. As a result, the crevasse stream leaves a deposit of sand (splay) that fans out across the floodplain from the levee breach.

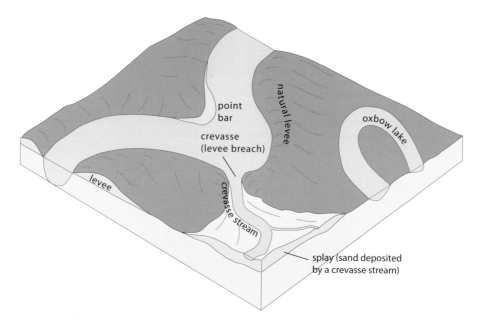

When the natural levee is breached during a flood event, a splay forms; if the breach is large enough, a new crevasse stream channel can form.

Numerous small rivers, including the famous Yazoo River, now occupy the Mississippi River Alluvial Plain. The term *yazoo stream* is actually derived from the Yazoo River. A yazoo stream is a tributary that lies within the floodplain of the main channel of a larger river and parallels it for a considerable distance before the two waterways join. The natural levee along the larger river generally prevents the two waterways from joining farther upstream.

Oxbow lakes are numerous in the Delta as well. Oxbow lakes form when pronounced meanders erode toward one another and actually join. When this happens, the river channel is shortened and the meander is abandoned, forming an arc-shaped standing body of water—an oxbow lake. The size of the oxbow reflects the size of the stream or river that formed it. In aerial view, the scars of hundreds of oxbow lakes that have been filled by natural processes are

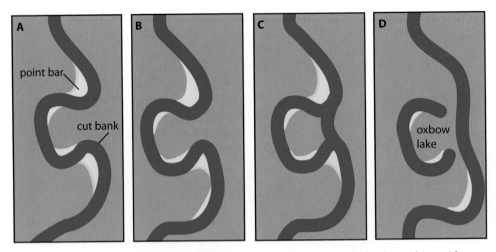

The formation of an oxbow lake. The river or stream cuts away at its banks on the outside of meanders (A and B), where the stream flow is greatest, creating cut banks. The main stem eventually cuts through the meanders (C), joining them. As the river deposits sediment, it abandons (D) the former meander, and the oxbow lake is born.

visible in the Delta. The ancestral Mississippi and Ohio Rivers formed the largest oxbow lakes in the Delta, including the five largest: Horn Lake, Moon Lake, Lake Bolivar, Lake Washington, and Eagle Lake.

HUMANS AND THE MISSISSIPPI

Because of the Mississippi River's flooding, fertile soils have developed on its floodplain. The soils of the Mississippi River Alluvial Plain are amongst the most fertile in the United States. However, the history of the alluvial plain shows that human settlement and floodplains do not mix.

Historical reports of flooding on the Mississippi date back to 1543, as noted during the Hernando de Soto exploration of North America. The river left its channel for eighty days. Since people first settled near the banks of the Mississippi River they have tried to limit flooding by constructing artificial levees. Unfortunately, as more levees were built and more of the river was confined, the higher the river rose in successive floods. Floods in the Delta were common, and by 1820 the federal government became involved, but its focus was navigation, not flood control. The flood events of 1849–50 received national attention and stimulated some flood control efforts.

Studies in the 1850s suggested three ways to limit flooding on the Mississippi: cutting through meanders to straighten the river, diverting tributaries to create artificial reservoirs and outlets, and constructing levees to confine the river. The first two solutions were deemed too expensive. In 1874 an act of Congress established the Mississippi River Commission to reclaim the Lower Mississippi River Alluvial Plain, with a focus on building levees. The levees that were in place at the end of the Civil War were in disrepair, and a major flood in 1879 destroyed them. Thomas Gregory Dabney was appointed engineer in charge in 1884, and he oversaw the construction of 100 miles of levees in Mississippi with a price tag of $15 million.

Humans entered the battle against the Mississippi River with force after the Great Mississippi Flood of 1927, which flooded 26,000 square miles and prompted national legislation and renewed funding to control the river. During the 1927 event the river was above flood stage for nearly six months in some areas and led to two hundred deaths and the evacuation of as many as six hundred thousand people. The 1928 Flood Control Act committed the United States to controlling the Mississippi and its tributaries. The plan called for additional levees and floodways and for channel improvement and stabilization.

The US Army Corps of Engineers was tasked with building a modern levee system large enough to limit the flooding of agricultural land. Undoubtedly, harnessing the Mississippi River in order to maintain channel integrity and to control flooding has been one of the greatest challenges undertaken by the corps. There are currently 2,203 miles of levees within the Lower Mississippi watershed, which includes all tributaries from Cairo, Illinois, south. Just over 1,600 miles of these levees are along the main Mississippi River channel, not its tributaries.

Once the corps started to tame the river, there was no turning back. The levees are working in Mississippi, but there are tradeoffs. A river forced to stay in its channel in one area will simply find another place to flood. Evidence of this was apparent during the 2011 flood event. In both Natchez and Vicksburg the volume and height of water in the Mississippi were greater than during the 1927 flood, but the total acreage in the Delta that flooded was less than half. While the levees did their job in Natchez and Vicksburg, the corps had to blow up a section of the levee in Missouri to let excess water flow into the Bird's Point–New Madrid Floodway.

Congress recognized that levee construction isolated many rivers and streams that once drained into the Mississippi River. The Flood Control Act of 1936 was passed to alleviate this problem. In Mississippi, the Yazoo is one of these isolated rivers. The purpose of the Yazoo Backwater Project, started in 1986, is to prevent backwater flooding of the Yazoo Basin. When the Mississippi River rises high enough to inhibit flow from the basin, levees and floodgates prevent the Mississippi from flooding into the Yazoo Basin, which comprises much of the southern portion of the Mississippi River Alluvial Plain south of Mississippi 12.

Although engineered levees constrain the Mississippi River in its channels, it does occasionally flood and renourish alluvial soils. The battle to control the river requires constant vigilance by the corps. If the Mississippi River had its way, it would be flowing south of Natchez through the Atchafalaya Swamp to Morgan City, Louisiana.

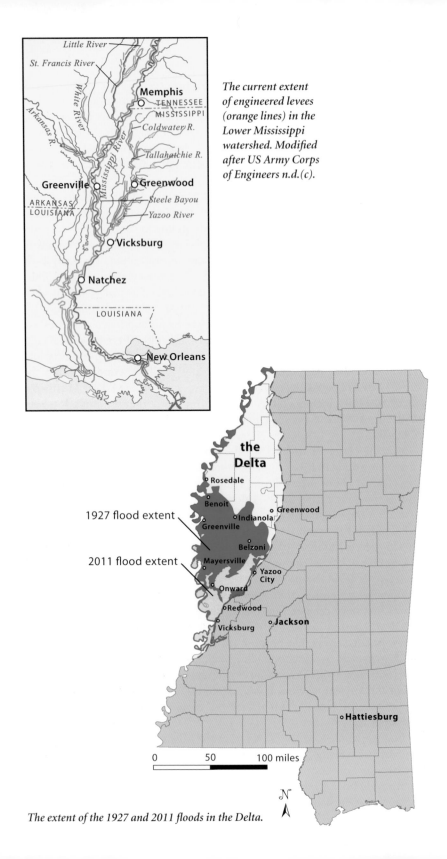

The current extent of engineered levees (orange lines) in the Lower Mississippi watershed. Modified after US Army Corps of Engineers n.d.(c).

The extent of the 1927 and 2011 floods in the Delta.

THE DELTA'S ECONOMIC AND SCIENTIFIC VALUE

Sand and gravel deposits are valuable commodities to the Mississippi economy. They are Mississippi's most valuable industrial minerals. More than seven hundred mines operate in the state, in nearly every county. Mines exploit modern gravel and pre-loess alluvial gravel likely deposited by the ancestral Mississippi, Ohio, and Tennessee Rivers, all three of which flowed through the state between 600,000 and 10,000 years ago.

The gravel deposits also yield important geological information. By identifying distinctive rocks and fossils in the gravel, geologists have been able to trace the history of the three rivers. Distinctive gravel specimens include Sioux Quartzite, the source rock of which is in southwestern Minnesota; Baraboo Quartzite, from Wisconsin; and Keokuk Quartz geodes, from Iowa, to name a few. These samples clearly indicate a source from the north. Glaciers plucked rocks from bedrock as they advanced southward, dumping their load when they melted. Cobbles and smaller particles were transported in the bed load of the ancient rivers, but others were much too large to have been moved by water. They must have been rafted to Mississippi, frozen in ice that broke free from glaciers. Many of the cobbles and boulders still bear scratches, or striations, that were made as the rock was dragged along bedrock at the base of a glacier.

The gravel also yields fossil shells of invertebrate organisms, such as corals, bryozoans, crinoids, trilobites, and brachiopods, that were originally deposited in Silurian-, Devonian-, and Mississippian-age limestones when ancient oceans still covered North America. The fossils were sourced from northern and eastern rock formations. For example, the ancestral Tennessee River brought the eastern fossils from the Appalachian Mountains. Petrified wood fragments are also abundant in pre-loess alluvial gravels.

The fossils and petrified wood were preserved through a process called *replacement*, in which silicon-rich groundwater replaces the original wood and shell material, creating the mineral chert. Chert has the same formula as quartz: silicon dioxide. Chert is very resistant to weathering, and it preserved the fossils during their long journey downriver. If the specimens had been their original softer composition of wood or calcium carbonate, they would not have survived.

The fertile soils in the Mississippi Delta have without a doubt played a more prominent role in the history of Mississippi than its gravels. Prior to the age of engineered levees, the Mississippi, Yazoo, Big Sunflower and other rivers deposited the fertile soils during floods. The majority of the Delta's fertile land is used for agricultural fields, pastures, and aquaculture operations. The soils can be grouped into two major types: meander belt or backswamp. The meander belt soils developed on point bars and natural levees. They are typically 9 to 16 feet thick and silty, sandy, and moderately well drained.

Backswamp soils, which developed on floodplains and in abandoned channels, are up to 30 feet thick and contain significant amounts of clay, which gives them a high capacity to shrink and swell based on water content. During periods of drought, wide vertical cracks, often more than 3 feet deep, develop in backswamp soils as the clay shrinks. With precipitation the clay swells and the cracks close up. Unless they also contain considerable sand, these soils do not drain well. This attribute was one of the leading factors in the growth and development of aquaculture in the Delta. The impermeability of the clay allows catfish producers to create ponds.

Catfish ponds are hard to miss in the Delta. In 2013 Mississippi led the nation in farm-raised catfish with nearly 50,000 acres under cultivation. The average yield is 5,000 pounds per acre. Catfish farming ranks fifth in revenue generated, placing behind cotton, corn, soybean, and eggs and poultry. Mississippi leads the nation in catfish production, though competition from imported catfish has led to a drop in the number of acres cultivated for aquaculture.

And finally, even though the Natchez Silt Loam constitutes less than 1 percent of the state's soils, it is the official state soil. This relatively young soil developed in the deep loess along the eastern margin of the Mississippi River Alluvial Plain and on the bluff of the Loess Hills physiographic province. If managed properly, it is suitable for crops and pastureland.

FROM THE ROAD

Roads in western Mississippi travel through the Mississippi River Alluvial Plain and the Loess Hills physiographic provinces. The alluvial plain dips gently toward the south and east, and elevations range from 200 feet at the state line near Memphis to 40 feet at Fort Adams, near the Louisiana state line at the very southwestern point of Mississippi. Beautiful vistas overlook the alluvial plain along numerous roads along the bluff line; in places the plain is up to 200 feet lower. Even though the Pleistocene- and Holocene-age alluvial deposits are relatively flat, elevation in the region is highly variable.

The alluvial deposits themselves are also highly variable. For example, oxbow-lake and backswamp deposits are typically fine-grained (silt and clay), whereas point bars, levees, crevasse splays, and channel deposits are composed of coarser-grained sand and gravel. Although rivers deposited most of the alluvial plain's sediment through channel migration and flooding, there are also alluvial fans composed of sediment eroded from the bluff line along the western edge of the Loess Hills physiographic province. The fans formed as high-gradient streams exited the bluff line onto the flat alluvial plain. The sudden change in gradient caused coarse particles to settle out and build up as a mound of sediment against the bluff. The topographic expression of the fans is often subtle and difficult to identify.

The Mississippi River Alluvial Plain contains numerous abandoned channels and courses; some are occupied by modern streams, but many are not. It is difficult to tell them apart, but in general length differentiates an abandoned channel from an abandoned course. Abandoned courses can be traced for tens of miles across the landscape, whereas channels cannot. Once water stops flowing in a course or channel, sediment begins to accumulate, filling them over several thousand years. Across the alluvial plain, courses and channels that aren't occupied by modern streams may still be noted by having slightly lower elevations than surrounding terrain, and they are easily identified by bald cypress (*Taxodium distichum*) and water tupelo (*Nyssa aquatica*), trees that thrive in them. After a recent rainfall, abandoned channels and courses often contain standing water.

The Mississippi River Alluvial Plain also has what's called *ridge and swale topography*, which develops as a river with a low gradient meanders across its valley. The most noticeable ridge and swale topography forms when channel migration is sporadic, meaning there is enough time for a good point bar to form before the channel again migrates to a new position and establishes a new point bar. When viewed from the ground, the subtle ridge and swale features are often obscured by agricultural practices, but aerial photographs reveal the pattern in barren fields quite readily.

Features of the Mississippi River Alluvial Plain are subtle, but there are generalizations about its geology that can be made. There are three main meander belts that were created over the past 10,000 years. The youngest is close to the present-day channel, with meanders that are up to 4,800 years old. The second is visible running through the center of the plain, with ages that range between 3,000 and 7,500 years. The third lies to the east, against the bluff line, with ages

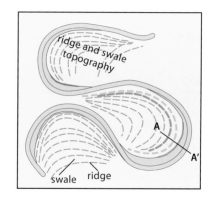

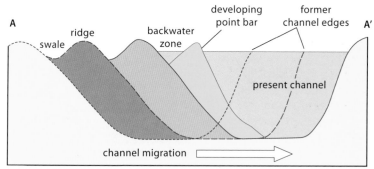

Ridge and swale topography. The line of cross section A-A' (top) crosses a river meander that is actively eroding and migrating to the southeast. The cross section shows the resultant ridges and swales (bottom).

that range between 2,600 and 10,000 years. As is evident from this pattern, over time the Mississippi has migrated westward. However, as recently as 2,600 years ago it occupied two main channels at the same time, one of which occurred in the most easterly of the meander belts. That's why it has such a wide range of ages.

Amongst the meander belts are Pleistocene-age braided-stream deposits that were laid down on the alluvial plain during the last glacial stage, which peaked approximately 22,000 years ago. A braided-stream channel is very different from the meandering channels present in the valley today. They develop when stream flow is intermittent, and periods of high discharge are followed by periods of low flow. As the discharge decreases, the river loses its ability to carry sediment, and it drops it in the channel, clogging it up. As a result, multiple small channels form that cross and flow around numerous sand and gravel bars that develop from the dumped sediment.

Two areas of Pleistocene-age sediment remain, one in the western portion of the Delta and another against the bluff line. There are also two low-lying areas of backswamp deposits, composed of silt and clay, that are observable in the southern and east-central parts of the alluvial plain.

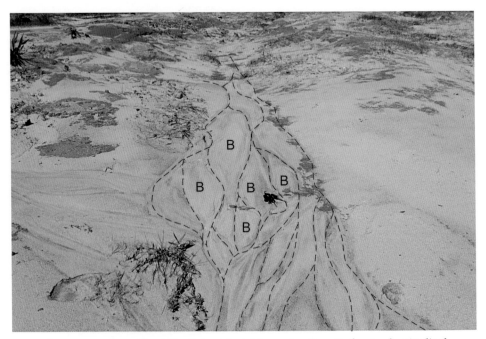

Great surges of runoff quickly dissipated at this construction site, leaving longitudinal bars (B) and multiple channels (dashed lines) in the sand that resemble those of braided streams. Huge volumes of glacial meltwater left much larger deposits on the Mississippi River Alluvial Plain during Pleistocene time.

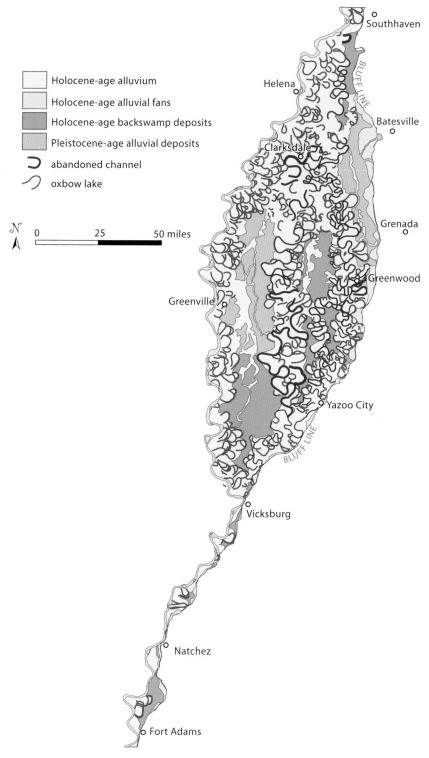

Map of Pleistocene- and Holocene-age alluvial deposits of the
Mississippi River Alluvial Plain. (Modified after Saucier 1994.)

Holocene-age alluvium
Holocene-age alluvial fans
Holocene-age backswamp deposits
Pleistocene-age alluvial deposits
abandoned channel
oxbow lake

N

0   25   50 miles

Southhaven

BLUFF LINE

Helena

Batesville

Clarksdale

Grenada

Greenwood

Greenville

Yazoo City

BLUFF LINE

Vicksburg

Natchez

Fort Adams

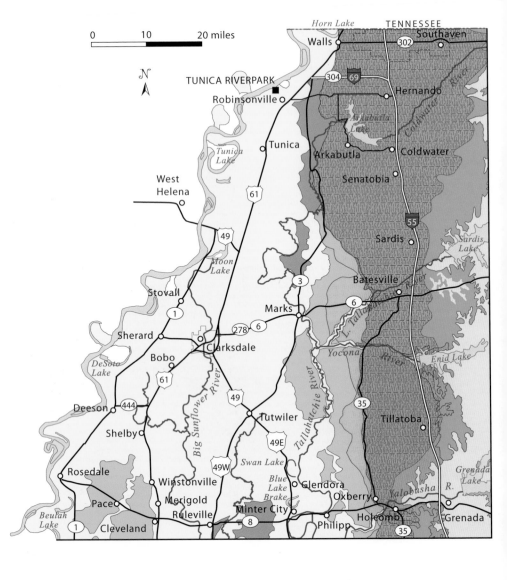

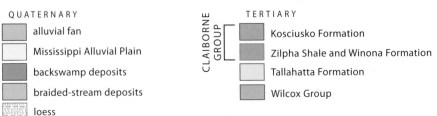

Geology along I-55 between Tillatoba and the Tennessee state line.

INTERSTATE 55
TILLATOBA—TENNESSEE
71 miles

The entire stretch of I-55 between Tillatoba and the Tennessee state line traverses the Loess Hills physiographic province. The hills of windblown silt cap pre-loess terrace alluvium and sand and clay of the Eocene-age Claiborne Group, deposited in deltas, nearshore environments, and deeper water during a period of elevated sea level.

From about mile marker 222 northward there are occasional small exposures of pre-loess alluvial gravel. Interstate 55 also crosses several rivers with younger floodplain and terrace alluvium. The more notable deposits are those of the Yocona River, flowing from Enid Lake near mile marker 232; and the Little Tallahatchie River, flowing from Sardis Lake near mile marker 244; along with Hickahala Creek (mile marker 266), Coldwater River (mile marker 273), and Hurricane Creek (just north of exit 283), all of which are tributaries of Arkabutla Lake. There are notable exposures of pre-loess terrace alluvium at Arkabutla Lake. (Take exit 271 at Coldwater or exit 280 at Hernando to access the lake. Also, see the Mississippi 304 (Scenic Route): Coldwater—Hernando road log for details.)

INTERSTATE 69/MISSISSIPPI 304
US 61—I-55
15 miles
See the map on page 122.

Interstates with odd numbers are generally routed north to south, but I-69 runs east to west, connecting US 61 with I-55. For about 2.5 miles from US 61, I-69 travels on sandy point bar deposits of Holocene-age alluvium, part of the Mississippi River Alluvial Plain physiographic province. I-69 then travels up the bluff marking the western edge of the Loess Hills physiographic province; it

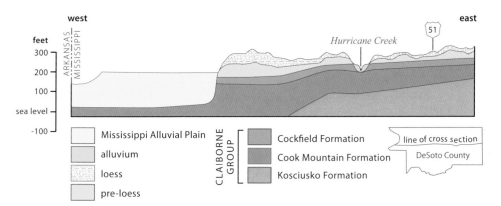

Cross section from the Mississippi River to US 51, just west of I-55, and paralleling I-69 just to the north. (Modified from Seal 1985.)

Looking eastward at the Loess Hills bluff from I-69.

follows these loess deposits, blown into place during Pleistocene time, to I-55. The hills cap pre-loess terrace alluvium and Claiborne Group sand of Eocene age. From mile marker 9 to its junction with I-55, I-69 crosses Hurricane Creek and its floodplain and terrace alluvium.

US 49E
Minter City—Tutwiler
22 miles

Between Minter City and Tutwiler are two abandoned channels of either the ancestral Mississippi or Ohio River. Swan Lake, just north of Glendora, occupies one on the east side of US 49E, and the Quiver River occupies the other south of Tutwiler.

One mile north of MS 8, US 49E crosses the southern end of Blue Lake Brake, a large (6 miles long and up to 0.6 mile wide), long-abandoned oxbow lake. Though not visible from the road, there is a sandbar present in the channel that developed in the river at the time the oxbow was cut off. The elevation of the sandbar relative to the surrounding channel has prevented bald cypress and water tupelo from growing on it. Local farmers take advantage of every inch of the Mississippi River Alluvial Plain that maintains slightly positive relief. The bar is nearly 4,000 feet long and 800 feet wide and provides about 40 acres of land suitable for planting. For comparison, Swan Lake, 6 miles to the north, is nearly the same length but less than 0.2 mile wide. The dimensions of an oxbow, channel, or course provide geomorphologists with important details about the size of the river that created the feature.

US 49W/US 49

RULEVILLE—TUTWILER—
CLARKSDALE—ARKANSAS

40 miles
See the map on page 122.

Between Ruleville and the Arkansas border, US 49W/US 49 travels on the Mississippi River Alluvial Plain physiographic province. Most of the route travels across sandy point-bar deposits of the central meander belt, with features ranging between 3,000 and 7,500 years old, except where it twice crosses an abandoned course of the ancestral Mississippi and Ohio Rivers: Hopson Bayou, just north of the MS 3 junction at Tutwiler, and Home Cypress Lake, 5.5 miles north of the junction.

Some abandoned courses become the channels of modern streams, and some do not. The old channel deposits feature coarse sediment, sand, and possibly gravel beneath finer-grained sediment that was deposited after the channel was abandoned. If occupied, modern streams are not likely to erode very deeply into the older deposits. If unoccupied, the channels will gradually fill in and disappear.

Rice farming about 0.5 mile north of Drew on the east side of US 49W. The linear features that rise slightly above the tops of the rice plants are earthen levees that the farmers sculpt around the perimeter of the field in the spring so the fields can be flooded to suppress weeds. The high clay content enables the temporary levees to retain their shape and hold water.

Looking east at Home Cypress Lake from US 49, just north of Dublin. The lake occupies an abandoned course of the ancestral Mississippi or Ohio River. The bald cypress trees in this 0.25-mile-wide channel are not dying. The image was taken in the fall, when the trees' foliage turns a beautiful orange-brown color before falling off. Bald cypress trees are deciduous conifers, which means they drop all their leaves for the winter, leaving bare branches. That is how they got their name.

Ridge and swale topography west of US 45, immediately north of MS 316, in a cotton field ready to be harvested. Ridge and swale topography is often very subtle. The arrows indicate the position of the ridges and swales.

At the MS 316 junction, about 12 miles north of the US 61 junction south of Clarksdale, ridge and swale topography is visible in the point bar deposits west of the highway. It is also visible 1 mile north of the MS 1 junction on both sides of the road (just east of the Mississippi River). Ridge and swale topography develops as a river with a low gradient migrates across the floodplain, leaving behind a series of point bars that creates the undulating terrain.

Also of note is the engineered levee of the Mississippi River that is visible about 1 mile north of the junction with MS 1. The current levee system has undergone continual reengineering and design since its inception. The more the river has been confined, the higher the floodwater level reaches in successive floods. As the river is constricted more and more to its channel, its water is less able to spread out onto the floodplain. A levee constructed in the late 1800s that was 9 feet tall and 53 feet wide now needs to be more than 30 feet tall and 315 feet wide in order to function.

The Mississippi's levees are constructed from clay-rich soils found along the banks of the river. They are compacted to minimize the movement of water through the levee. Even so, levees are not absolutely impervious, so engineers must carefully calculate the pressure generated by the river's water as it becomes elevated during a flood. As the height of the floodwater increases, the pressure on the levee increases, and the velocity of the water flowing through the earthen levees increases. If the velocity gets too great, the levee can fail as it is carried away grain by grain.

Water will always take the path of least resistance. There are several ways for a levee to fail. In addition to simply being overtopped by floodwater, levees can fail by underseepage and piping. Underseepage occurs when water seeps out of the channel through the material beneath the levee. Once it reaches the landward side of the levee, the pressure of the elevated floodwater quickly washes away the underlying material and the levee itself. Piping is similar, but it happens when water flows through the levee and erodes levee material from the river side landward. Once the hole works its way through the levee, the river quickly enlarges the hole and causes failure. In any of these scenarios, as much as 2,000 feet of levee can collapse in a few seconds.

US 61
CLEVELAND—CLARKSDALE—TENNESSEE
100 miles
See the map on page 122.

From Cleveland to Walls, US 61 traverses the Mississippi River Alluvial Plain physiographic province. Pleistocene-age braided-stream deposits, remnants of the deposition that took place on the alluvial plain during the last ice age, which peaked about 22,000 years ago, lie immediately west of the highway at Cleveland. Otherwise, US 61 crosses the Holocene-age meander belts of the Mississippi River.

Although not distinguishable from the road, 11 miles north of the junction with MS 8, at Winstonville, US 61 crosses from the western meander belt to the central meander belt. Then, farther north, at Clarksdale, the road crosses the boundary between the central and eastern meander belts. Generally speaking, the Mississippi occupied the eastern meander belt first and migrated westward over time. The youngest meander belt is close to the present-day channel, with deposits ranging up to 4,800 years old. The central belt trends through the center of the plain and has deposits ranging between 3,000 and 7,500 years old. The eastern belt lies against the bluff and has deposits with a wide range of ages, between 2,600 and 10,000 years old. The wide range of dates for the eastern meander belt is because the Mississippi occupied two channels as recently as 2,600 years ago.

Abandoned channels of the Mississippi and Ohio Rivers, some containing water, are everywhere. Notable examples include, from south to north, Jones Bayou, north of Cleveland; Little Mount Bayou, north of Merigold; the Hushpuckena River, 4 miles north of MS 32 at Shelby; Harris Bayou, just north of Bobo; the Big Sunflower River at Clarksdale; Swan Lake on the east side of US 61, 6 miles north of the MS 6 junction at Clarksdale; Yazoo Pass, 13 miles north of the MS 6 junction; Ark Bayou, 8 miles north of the US 49 junction; Beaverdam Bayou, 10 miles north of the US 49 junction; White Oak Bayou, at the MS 4 junction at Evansville; Johnson Creek, about 4 miles north of the MS 3 junction; and Horn Lake north of Walls.

Tunica RiverPark

Tunica RiverPark, west of Robinsonville, contains a museum with many exhibits about the Mississippi River, including a movie about the 1927 flood, the technology used to control the riverbanks, the soils of the alluvial plain, and wetlands. Two relatively short trails feature great perspectives of the Mississippi River and traverse two wetlands that have developed in abandoned channels. Riverboat cruises are also available.

As noted already, the Mississippi River floods, sometimes disastrously. The frequency at which natural disasters occur is inversely related to their magnitude, and the greater the magnitude of the event, the less frequently it is likely to occur. For the majority of the year, the Mississippi stays in its channel. Throughout the basin during springtime, however, snowmelt and seasonal rain bring more water to the river, and the river's level rises. Eventually it will reach flood stage and leave its channel to flow out onto the floodplain. Flood stage may be reached several times a year, but little to no damage usually occurs. Severe floods, which, by definition, affect larger swaths of land, require more water, so they occur less frequently.

Using historical data, the discharge values of a river can be ranked from the greatest to the least. The discharge value is the volume of water flowing past a given point in a given amount of time, usually measured in cubic feet per second. Most of the time the Mississippi flows at intermediate discharge values, but it also experiences peak and minimal flows each year. Based on historical data averages, if a river achieves a peak discharge value 1 percent of the time, then there is a one in one hundred chance that this flow rate will occur in any given year. Scientists refer to this as the *recurrence interval*, which is basically a measure of how often this peak flow is likely to occur; however, it is popularly known as the "one-hundred-year flood." Likewise, if there is a 2 percent chance of a particular peak discharge occurring in any given year, this flow level is said to have a fifty-year recurrence interval—also called the "fifty-year flood."

There is a display south of the RiverPark Museum with markers representing water levels for five-, ten-, twenty-five-, fifty-, and one-hundred-year flood levels at this location. The level of the bank here is 20.3 feet above the mean water level of the river. There is a 20 percent chance that water will reach the elevation of the bank— the five-year flood level—in any given year. The 1927 flood would have covered this location with 11 feet of water. Based on historical discharge levels, the probability of that flood level happening again in any year is 2 percent, a recurrence interval of fifty years. The 2011 flood was 2.45 feet higher and was classified with a recurrence interval of one hundred years.

Monuments marking the heights of several historic Mississippi River floods at Tunica RiverPark. Ground level represents a recurrence interval of five years (five-year flood); heading counterclockwise, the markers represent recurrence intervals of ten, twenty-five, fifty, and one hundred years. The tallest marker is the height of the levee that was constructed at Tunica.

Revetments are popular structures for maintaining a river's channel position. Revetments can be wood, rock, or concrete placed on the sloping sides of a channel to prevent erosion, and therefore channel migration. Early in the 1900s, revetments on the Mississippi were made from willow branches woven together and then sunk. Modern revetments are made of riprap (loose piled boulders) or cast concrete beams that are linked to one another on a barge and then laid on the banks of the river. At Tunica RiverPark, riprap composed of limestone boulders lines the banks of the river.

Mile 0 of the Mississippi River's engineered levee on the west side of US 61. The plaque on the monument to the left of the marker, dedicated to Thomas Gregory Dabney, reads, "To his broad vision and unswerving purpose the development and the security of this alluvial empire is in large part due."

Though the topographic expression is very subtle, there is an example of ridge and swale topography immediately north of the MS 315 junction, on the west side of US 61. Here the axes of the ridges and swales are oriented northwest-southeast.

The Tunica Museum contains numerous hands-on displays about the history of Mississippi River and offers nature trails through wetlands. It's about 3 miles north of the MS 4 junction in Tunica, about 1 mile west of US 61 off Perry Road.

US 61 travels up and into the Loess Hills physiographic province north of Walls. About 1.5 miles north of the MS 302 junction, US 61 crosses the engineered levee of the Mississippi River. Thomas Gregory Dabney was appointed the engineer in charge of reining in the Mississippi in 1884. He oversaw the construction of 100 miles of levees in Mississippi for $15 million. A marker stands as a tribute to his work at mile 0 of the levee system.

US 278/MISSISSIPPI 6
Clarksdale—Batesville—I-55
37 miles
See the map on page 122.

Except where it crosses abandoned courses of the ancestral Mississippi and Ohio Rivers, US 278/MS 6 traverses point bar deposits from Clarksdale to about 1 mile east of the MS 3 junction at Marks. A swamp on the north side of the road occupies an abandoned channel about 3.5 miles east of the US 49 junction, at Clarksdale. Modern streams or water bodies occupy several abandoned courses and channels, including Big Creek Deadwater, Cassidy Bayou, and the Coldwater River. The Coldwater River occupies a former distributary of the Mississippi.

Near Clarksdale the alluvial sediment is between 3,800 and 6,200 years old but increases to between 9,000 and 10,000 years old near Marks. From there the road crosses Pleistocene-age braided-stream deposits for about 10 miles. These deposits were laid down along the Mississippi floodplain during the last glacial stage. A braided-stream channel is very different from the meandering channels present in the alluvial valley today. A braided channel forms when stream flow is intermittent and periods of high discharge are followed by periods of low flow. As the discharge decreases, the river loses its ability to carry sediment and it drops it in the channel. The result is multiple small channels crossing and flowing around numerous sand and gravel bars. There is no change in elevation across these Pleistocene deposits, but in aerial photos the scars left behind by braided streams are distinctive, looking like braided hair.

After crossing the Little Tallahatchie River floodplain, US 278/MS 6 traverses alluvial and point bar deposits of the ancestral Mississippi and Ohio Rivers before leaving the Mississippi River Alluvial Plain physiographic province. The change in elevation is not as dramatic at Batesville as in other locations along the bluff line because the highway was constructed on the southeastern side of the Little Tallahatchie River floodplain, where the river eroded the bluffs in the past. From its junction with MS 35 at Batesville, the road travels in the Loess Hills physiographic province to I-55. The loess caps pre-loess terrace gravel and sand and the Eocene-age Kosciusko Formation. The loess thins out and disappears about 4 miles east of I-55.

MISSISSIPPI I (GREAT RIVER ROAD)
ROSEDALE—US 49
54 miles
See the map on page 122.

Between Rosedale and US 49, MS 1 travels across sandy point-bar deposits, the abandoned channels of the Mississippi and Ohio Rivers, or abandoned courses of their crevasse streams and distributaries. It also traverses the western meander belt of the modern Mississippi, with channels and deposits ranging up to 4,800 years old. Modern rivers, streams, and bodies of water occupy the ancient and more modern channels and courses. The most notable include Bogue Phalia, 1.5 miles north of the MS 32 junction; a swamp about 3.5 miles north of the MS 444 junction, north of Deeson; Richies Bayou, 2 miles north of the MS 322 junction at Sherard; the Big Sunflower River, about 3 miles north of Stovall; Long Lake, to the east, about 6 miles north of Stovall; and Moon Lake, about 3 miles north of Long Lake.

About 2 miles north of the MS 32 junction, ridge and swale topography of point bar deposits is visible on both sides of the highway. Ridge and swale

Yazoo Pass

In February 1863, as part of the Yazoo Pass Expedition, Union forces intentionally blasted the Mississippi River's levee to flood Moon Lake. Their plan was to use the additional water to float a flotilla to Vicksburg via Yazoo Pass and the Coldwater, Tallahatchie, and Yazoo Rivers. Although the plan worked, their progress was slow enough that the Confederate Army was ready for them when they reached Fort Pemberton, in Greenwood, in April. A historical marker on MS 1, about 3.5 miles north of the turnoff to Moon Lake, commemorates this wartime effort.

topography develops as a river with a low gradient widens its valley, leaving behind a series of point bars that creates the undulating terrain. The elevation difference between the ridge and swales is less than 10 feet, but the distance between ridges may be up to 100 feet.

The Delta is one of the leading agricultural areas in the United States. Mississippi ranks first in catfish production; third in pulpwood, sweet potatoes, cotton, and cottonseed; fifth in rice; and twelfth in soybeans, to name a few of the top-ranking commodities, most of which are grown in the Delta. What Mississippi lacks in mineral resources it makes up for agriculturally with its rich alluvial soils and the Mississippi River Valley Alluvial Aquifer. Of all the counties in the Delta, Bolivar County is one of the most productive; it leads Mississippi in total acres for rice, soybean, and wheat production. It is no surprise, then, that this county also uses the most water for irrigation—25 percent more than second-place Sunflower County. Mississippi and Louisiana both receive approximately 60 inches of precipitation annually, the highest averages of the lower forty-eight states and second only to Hawaii. Despite all that precipitation, crops still require irrigation due to the excessive amount of moisture lost to evaporation.

MISSISSIPPI 8
Rosedale—Cleveland—Grenada
73 miles
See the map on page 122.

From the MS 1 junction at Rosedale to the MS 35N junction, MS 8 travels through the Mississippi River Alluvial Plain physiographic province. The 60-mile trek crosses the widest expanse of this province in the state, passing through all the important regions of the Mississippi River Alluvial Plain, including three major meander belts that increase in age from west to east, two Pleistocene-age braided-stream deposits, backswamp deposits, alluvial fans that developed off the bluff line, and pre-loess terrace alluvium and loess in the bluffs.

For about 6 miles east of MS 1, MS 8 travels on point bar deposits, except where it crosses abandoned channels of the ancestral Mississippi and Ohio Rivers and the abandoned course of a crevasse stream, or distributary, of these rivers. Then the road passes over Pleistocene-age braided-stream deposits to Jones Bayou, just west of US 61/US 278 junction in Cleveland.

About 2 miles east of Pace the highway crosses the straight, engineered channel of Snake Creek; 1.5 miles later it crosses the natural Snake Creek channel. The engineered cutoff, just under 1 mile in length, shortens the channel by 3.4 miles. The channel was most likely cut to convey water more efficiently downstream and to limit localized flooding of the farm just off MS 8. Without a constant supply of water the former channel will gradually fill in with vegetation and sediment.

For about 16 miles east of Cleveland, MS 8 travels again on point bar deposits and abandoned channels of the ancestral Ohio and Mississippi Rivers, which are occupied by Jones Bayou at Cleveland, Dougherty Bayou north of Ruleville, and the Quiver River east of Ruleville. The road then transitions onto clayey backswamp deposits for the next 6 miles, to roughly the turnoff to Schlater (County Road 522). From County Road 522 to about 6 miles east of the US 49E junction at Minter City, the road again travels on point bar deposits.

At Minter City, MS 8 crosses an abandoned course occupied by the Old Channel of the Tallahatchie River. A new channel was cut less than 5 miles east of the US 49E junction to shorten the flow path and decrease flooding potential. East of the Old Channel crossing, MS 8 parallels the Tallahatchie River for about 7.5 miles.

From Philipp to the MS 35N junction at Oxberry, MS 8 travels again on Pleistocene-age braided-stream deposits, except where it crosses point bar deposits of several small abandoned streams, east of Philipp, and at the bluff line, where alluvial fans overlie the Pleistocene-age sediment. The alluvial fans are fan-shaped features that have developed where small streams erode sediment from the bluff line and deposit it against the bluffs. The sudden change in stream gradient and loss of channel confinement causes the streams to drop their sediment load.

From Oxberry to the I-55 junction at Grenada, MS 8 travels in the Loess Hills physiographic province. However, much of this stretch of road is on the floodplain and terrace alluvium of the Yalobusha River and its tributaries. MS 8 traverses a couple of loess-capped hills about 1 and 6 miles east of the MS 35S junction at Holcomb. The hills are underlain by pre-loess terrace sand and gravel and sand and clay of the Eocene-age Claiborne Group.

MISSISSIPPI 304 (SCENIC ROUTE)
COLDWATER—HERNANDO
31 miles

MS 304 Scenic Route (SR) travels around Arkabutla Lake. It starts at exit 271, at Coldwater, as Arkabutla Road and then joins MS 301 at Arkabutla, heading north and connecting with MS 304SR through Hernando, connecting back to I-55 at exit 280. This route and Arkabutla Lake fall in the Loess Hills physiographic province.

For about 3.5 miles west from exit 271, MS 304SR travels on the floodplain and terrace alluvium of both the Coldwater River and Hickahala Creek. The Hickahala Creek deposits are west of the Coldwater River terraces. The road also crosses deposits of the Coldwater River about 4 miles north of Arkabutla.

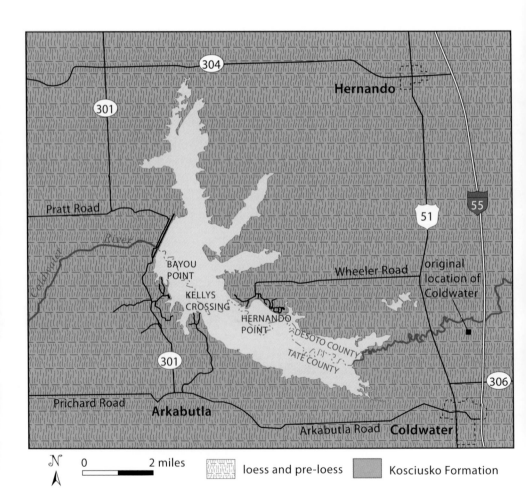

Geology of the Arkabutla Lake region.

A notable rise in elevation, where the road passes from the modern floodplain to the youngest terrace, occurs about 4.5 miles north of Arkabutla. The Coldwater River flows through a former distributary of the Mississippi. The channel is estimated to be between 9,000 and 10,000 years old. In the Mississippi River Alluvial Plain there are only a few places where deposits this old remain.

Besides these floodplains and terraces, the rest of the route travels through hills of windblown silt that cap pre-loess terrace sand and gravel. There are multiple exposures of loess along this route, but notable exposures are found on the west side of MS 304SR about 8.5 miles north of Arkabutla, and about 6 miles east of the MS 301/304 junction. The pre-loess deposits are a valuable economic commodity in the Memphis metropolitan area. Although sand and gravel are not expensive resources to obtain, their proximity to areas of use is critical to keep transportation costs low.

Arkabutla Lake

Ninety-square-mile Arkabutla Lake is impounded behind a 2.2-mile-long dam across the Coldwater River. In 1942 the US government moved the seven hundred inhabitants of Coldwater 1 mile south to accommodate the project. There are notable exposures of loess and pre-loess terrace alluvium at Kellys Crossing (20 feet) and Hernando Point (30 feet). At Bayou Point there is an excellent outcrop of pre-loess alluvium that has been cemented by iron, forming sandstone and conglomerate. This material was deposited sometime after the Eocene-age Kosciusko Formation but before the loess,

About 20 feet of loess exposed at Kellys Crossing. The vertical loess bluff breaks at the contact with pre-loess sand and gravel. Iron-cemented pre-loess sand and gravel are visible above the waterline.

The vertical loess bluff above pre-loess sand and gravel at Bayou Point. The ledges in the foreground are iron-cemented pre-loess gravel, a rock called conglomerate.

Close-up of the conglomerate ledges at Bayou Point. The well-rounded pebbles in the conglomerate are primarily chert.

hence the designation *pre-loess*. The iron in the sandstone and conglomerate is a result of weathering. Chemical weathering of the original sand and gravel released iron in the minerals to groundwater. When the water was exposed to oxygen in the atmosphere, iron oxides and iron hydroxides precipitated out around the sand grains and gravel, cementing them together. The source of iron is not apparent. Some Tertiary-age rocks contain iron-rich glauconite, but the pre-loess terrace alluvium does not.

US 49E
Yazoo City—Greenwood—Minter City
74 miles

For about 2 miles from its split with US 49W at Yazoo City, US 49E travels through the Loess Hills physiographic province. The hills are made up of wind-blown silt of Pleistocene age that caps pre-loess alluvial gravel and the under-lying Eocene-age Jackson Group, primarily the Yazoo Clay. The highway then drops off the loess and travels along the Mississippi River Alluvial Plain phys-iographic province, traversing the eastern meander belt and the modern flood-plains of the Yazoo and Tallahatchie Rivers. The confluence of the Tallahatchie and Yalobusha Rivers, just north of Greenwood, forms the Yazoo River.

Many people mistakenly think that alluvial fans exist only in arid regions, but this is not correct. They are certainly more apparent in arid regions because vegetation is absent, but alluvial fans will develop wherever a stream experi-ences a sudden loss of confinement and a sudden change in gradient. When streams leave the loess hills and flow out onto the alluvial plain, fans form. Many of the alluvial fans along the bluff line are small and easy to miss. A slight rise in elevation just before a stream valley crossing indicates that you just crossed a fan. Though subtle changes in elevation on the alluvial plain seem unimportant, relief of as little as 1 foot can save property from flooding. Where alluvial fans and point bars are available, homes are generally built on them because they are the landforms with the greatest elevation. Every foot of flood-water rise requires substantially more water, so any advantage in elevation can be important when it comes to flooding.

Looking north at a spot where US 49E passes over the subtle rise of a small alluvial fan.

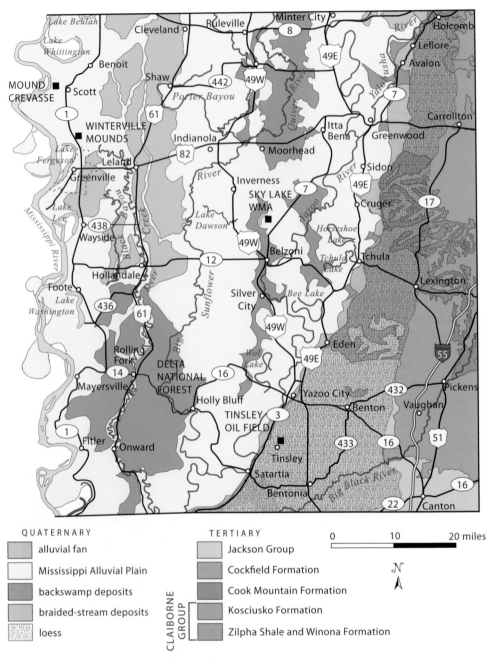

Geology along US 49E between Yazoo City and Minter City.

QUATERNARY

- alluvial fan
- Mississippi Alluvial Plain
- backswamp deposits
- braided-stream deposits
- loess

TERTIARY

- Jackson Group

CLAIBORNE GROUP
- Cockfield Formation
- Cook Mountain Formation
- Kosciusko Formation
- Zilpha Shale and Winona Formation

0   10   20 miles

N

From Eden to Minter City, US 49E travels on the Mississippi River Alluvial Plain, mostly sandy point-bar deposits of the Ohio and Mississippi Rivers. US 49E passes three abandoned courses of the ancestral Mississippi and Ohio Rivers: Bee Lake, about 13 miles north of the US 49/US 49W split; Horseshoe Lake, between Tchula and Cruger; and Roebuck Lake, just southwest of Greenwood. Because their shapes—about 170 yards wide and 15, 13, and 19 miles long respectively—are not typical of lakes in most states, they are excellent examples of abandoned courses.

Ridge and swale topography is visible on the east side of the road immediately north of Holly Grove-Meeks Road in Cruger, and on the west side of US 49E, immediately north of the Staplcotn warehouses in Rising Sun (south of Greenwood). Ridge and swale topography develops as a river meanders across the floodplain, leaving behind a series of point bars that create the undulating terrain. The elevation difference between a ridge and swale is less than 10 feet, and the peaks of the ridges occur at intervals of about 120 feet.

Ridge and swale topography north of Cruger.

Perspective from US 49E at Tchula, looking northeast across the Mississippi River Alluvial Plain at the bluff line of the Loess Hills, approximately 5 miles away.

US 49W
YAZOO CITY—BELZONI—
INDIANOLA—RULEVILLE
93 miles
See the map on page 138.

Yazoo City is known as the "Gateway to the Delta." US 49W originates just north of Yazoo City, where US 49 splits into eastern and western routes. US 49W travels through the Mississippi River Alluvial Plain physiographic province, traversing the easternmost edge of the central meander belt and the Sunflower River floodplain. The route is mostly across sandy point-bar deposits of the meander belt. However, there are exceptions: where US 49W crosses backswamp deposits, 1 mile north of the US 49W/49E split, between Silver City and Belzoni, and at Sunflower and Cottondale. While driving through the area, only the lack of channels and oxbow lakes indicates a backswamp area versus meander belt, but this is tough to notice at 60 miles per hour!

About 3.5 miles north of the US 49 split, US 49W crosses the Yazoo River, the other major river along this route that flows in abandoned channels and courses of the ancestral Mississippi and Ohio Rivers. The highway crosses the Big Sunflower River twice, just south of Indianola and again 3 miles north of US 82. After the second crossing, US 49W parallels the river for the rest of the trip into Ruleville.

Mississippi leads the United States in catfish production, and US 49W crosses through Sunflower County and the heart of catfish country. When it comes to catfish production, this county is second only to Leflore County, immediately to the east. It is hard not to notice the acres and acres of catfish ponds in the area. The geology of the Delta is a major reason for the success of the catfish industry in Mississippi. Much of the acreage under fish farms lies within a region of backswamp deposits resting between the central and eastern meander belts, along the Leflore-Sunflower county line. Catfish ponds are easily sculpted into the flat floodplain, the fine-grained clay creates a nearly impermeable bottom that helps the fishponds hold water, and water from the Mississippi River Valley Alluvial Aquifer is readily available. The domestic catfish industry is currently struggling, as imported foreign fish is flooding the market.

Low stream gradients and deep channels make it difficult to tell when water bodies are modern streams and rivers or lakes occupying old channels. In either case, much of the area surrounding the water is classified as wetlands. Wetlands are identified based on three criteria: (1) there must be water at or near the surface for a portion of the year; (2) there must be habitat for hydrophytes, water-loving trees and plants; and (3) the soils must reflect a long history of being water saturated. Based on these criteria, much of the Delta is or was considered wetland. More than 60 percent of the natural wetlands have disappeared from the region since Mississippi was settled. Once forested with timber, Mississippi's trees were cut and canals were dug into the wetlands to drain the soils and create agricultural fields. Despite levee construction, some areas remain susceptible to local flooding, which minimizes crop yield or destroys crops altogether. There are federal programs that buy back nonproductive acreage and let the land return to a natural state. Such programs not only increase wetland habitat but also reduce insurance claims in frequently flooded areas. An excellent example of a freshwater wetland is Sky Lake Wildlife Management Area.

Sky Lake Wildlife Management Area

The Sky Lake Wildlife Management Area, about 8 miles north of Belzoni off MS 7, encompasses more than 4,000 acres and features a 1,735-foot-long elevated boardwalk that leads into the heart of Sky Lake, a bald cypress–tupelo wetland. Sky Lake is an oxbow of the Mississippi River that was cut off from the main channel between 3,800 and 5,000 years ago. The lake is in the middle-age stage of its life. After its channel was cut off from the Mississippi, fine-grained silt and mud washed in from the surrounding landscape during flood events and began to accumulate. As more and more sediment accumulates, the lake bed fills in, but due to compaction it is still nearly 20 feet below the level of the old natural levee surrounding the original channel. The lake bed is often exposed in the summer, but the channel can fill during the late winter with as much as 16 feet of water.

Geologists now recognize that silt and mud can be contaminants in and of themselves, but they also have the ability to attract harmful chemicals and elements and carry them to rivers and streams. Wetlands can serve as natural filters, capturing sediments and sequestering the contaminants stuck to them. Clay traps some of the elements and chemicals, and the organic matter generated by the cypress and tupelo trees and other vegetation traps others. Bald cypress (*Taxodium distichum*) and water tupelo (*Nyssa aquatica*) are hydrophytes, trees that thrive when their roots are

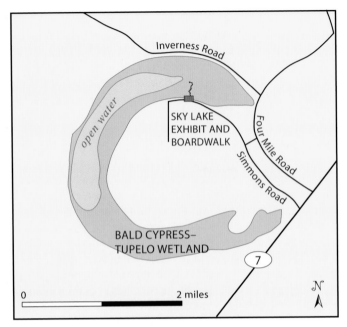

Sky Lake Wildlife Management Area.

Large bald cypress, as observed from the elevated walkway at Sky Lake during low water level. The water level can rise to 16 feet in the spring, which would place it at the base of the walkway.

This marker, just outside the channel of Sky Lake, shows the levels of several large floods of the Mississippi River. The 1927 flood was 9 feet above the ground here, which is more than 30 miles from the Mississippi River.

flooded, and they are a mainstay of freshwater wetlands like Sky Lake. Scientists have found the fingerprints of early inorganic pesticides, such as calcium and lead arsenate used in the 1920s to control the boll weevil, preserved in the sediment record at Sky Lake. So not only are wetlands magnificent places to behold, they also serve a critical role in preserving water quality in the Delta.

Logging and efforts to create arable cropland in the late eighteenth and nineteenth centuries resulted in the loss of more than 60 percent of the wetlands in Mississippi. The loss of natural vegetation and constant tilling of soil over the past century or so has resulted in a significant increase in the sedimentation rate in Sky Lake and other Delta lakes. They are filling in faster than at any time since they were cut off from the river. Investigations indicate that Sky Lake has experienced as much as a 300 percent increase in sediment accumulation since the late 1800s and early 1900s. Sky Lake may fill in sooner due to modern rates of erosion, but it will continue to trap and store harmful contaminants, as it has always done. Because wetlands are an important ecologic and economic resource, Mississippi River Alluvial Plain regulations administered by the US Environmental Protection Agency and US Army Corps of Engineers offer them some protection.

US 61
ONWARD—CLEVELAND
77 miles
See the map on page 138.

From Onward to Cleveland US 61 is entirely in the Mississippi River Alluvial Plain physiographic province, mostly passing through Holocene-age back-swamp and Pleistocene-age braided-stream deposits. US 61 traverses the division between the modern, or western, meander belt of the Mississippi and the central meander belt with deposits that are between 3,000 and 7,500 years old. Between Onward and Leland, the route parallels and often crosses Deer Creek, which occupies an abandoned channel of the ancestral Mississippi and Ohio Rivers.

Leroy Percy State Park is about 6 miles west of Hollandale, off MS 12. Black Bayou, within the park, occupies an abandoned course of a small crevasse stream or distributary of the Mississippi River. However, alligators are the park's main attraction. Except for a few tenths of a mile at Hollandale, where the road passes over Deer Creek and its alluvium, MS 12 travels over braided-stream deposits all the way to the park.

Looking east at Porter Bayou, which occupies an abandoned course of the Mississippi and Ohio Rivers, immediately north of the MS 448 junction in Shaw.

See the map on page 138.

US 82
Arkansas—Greenville—Indianola—Greenwood—Carrollton
82 miles

From the Mississippi River to beyond Greenwood, US 82 travels through the Mississippi River Alluvial Plain physiographic province. The route crosses the Mississippi state line just east of the bridge above the Mississippi River. Lake Chicot, immediately west of the state line, is an oxbow of the Mississippi that formed very recently. At 17 miles in length and just over 0.5 mile in width, it is one of the largest oxbows related to the Mississippi River. The western boundary of Greenville is Lake Ferguson, which is also a recently abandoned channel of the Mississippi River. The state line actually runs through the lake.

From the state line to about 1 mile east of the MS 1 junction at Greenville, US 82 travels on point bar deposits and abandoned channels of the western meander belt, with ages that range up to 3,000 years. From 1 mile east of Greenville for a distance of 2 miles, and then from 1 mile east of the crossing at Deer Creek for 6 miles, US 82 crosses Pleistocene-age braided-stream deposits that geoscientists have recently dated; they are just under 15,000 years old, having been deposited toward the end of the last Pleistocene glaciation.

Deer Creek, which US 82 crosses just west of Leland, may have formed as a crevasse stream off the Mississippi River during Holocene time. It flows to the southeast through the southern portion of the alluvial plain and was a tributary of the Yazoo River, just north of Redwood, prior to levee construction along the Yazoo. Due to the narrow width of its channel, it's unlikely that it carried much of the Mississippi's discharge.

Bogue Phalia, 2 miles east of the US 61 junction, is possibly another distributary of the Mississippi that formed from a crevasse just north of Rosedale. The Bogue Phalia occupies a former channel that had been eroded down through Pleistocene-age braided-stream deposits. About 5 miles east of the Bogue Phalia, US 82 crosses into the central meander belt, with channels, oxbows, and courses that formed between 3,800 and 7,500 years ago. The Big Sunflower River, about 4 miles east of US 49W, flows within this meander belt.

From its junction with MS 3 at Moorhead to the junction with CR 507, at Mississippi Valley State University, US 82 travels on Holocene-age backswamp deposits. Catfish aquaculture is big business in this part of the Delta. Two factors favor the construction of large catfish ponds: access to adequate water and relatively impermeable, clayey soils. While catfish ponds are visible over most of the Delta, Leflore County is the leader in production. The backswamp deposits in this region, eastern Sunflower and western Leflore Counties, are prime locations for catfish ponds. Several are visible along US 82 east of Moorhead. The circular cages on top of the water are aerators that oxygenate the water as temperatures rise in the summer and oxygen levels in the water drop.

From MS 507 to 1 mile east of the MS 7N junction the road crosses the point bars, abandoned channels, and courses of the eastern meander belt, remnants

Aquaculture, primarily catfish, is big business in the Mississippi Delta and is certainly related to geology. Artificial levees are easily formed in the flat clay-rich soil to create 4-to-6-foot-deep ponds, and the Mississippi River Valley Alluvial Aquifer provides a reliable source of water.

from when the Mississippi River was active in the region between 2,600 and 10,000 years ago. Roebuck Lake, south of Greenwood, and the Yazoo River both occupy former courses. The engineered cutoff just east of the US 49E junction shortened the length of the Yazoo River by 11 miles. The cutoff permits floodwater to flow past Greenwood, which sits in the center of a meander of the Yazoo, and minimizes flood impact. Just north of Greenwood is where the Yalobusha and Tallahatchie Rivers come together to form the Yazoo.

East of MS 7, US 82 crosses 2 miles of Pleistocene-age braided-stream deposits before starting to climb an alluvial fan. The fan, which developed off the bluff line during the Holocene, lies over the braided-stream deposits. It rises at a gradient of about 14 feet per mile to the bluff line, which is about 2.5 miles east of the Carroll County line. From there to Carrollton, US 82 travels in the Loess Hills physiographic province. The loess here caps Pleistocene-age pre-loess alluvial gravel and Kosciusko Formation sand. At the junction with MS 17 loess is thin to nonexistent.

Mississippi Alluvial Fans

Though they are often associated with arid climates, alluvial fans form in all climates. In cross section, fans are triangular-shaped deposits that develop where stream channels are no longer confined and they experience a sudden decrease in elevation, such as where one exits the mountains onto a plain. Besides a lack of confinement, streams must also have sufficient sediment and the strength to carry that sediment in order for a fan to develop. The head of the fan is its apex, where the stream is first unconfined. As the stream encounters flatter terrain it slows down, dropping its sediment load, which spreads out beneath the apex like an apron. In western Mississippi, fans form where stream channels that are deeply eroded into the loess and sediment of the bluff line flow onto the alluvial plain. They are called *humid* alluvial fans because they form in a humid climate.

The alluvial fans that form off the bluff line, from Yazoo City to Southaven, have been a target of research by the Mississippi Office of Geology in recent years. As water use in the Delta increases, there is a critical need to understand the mechanisms by

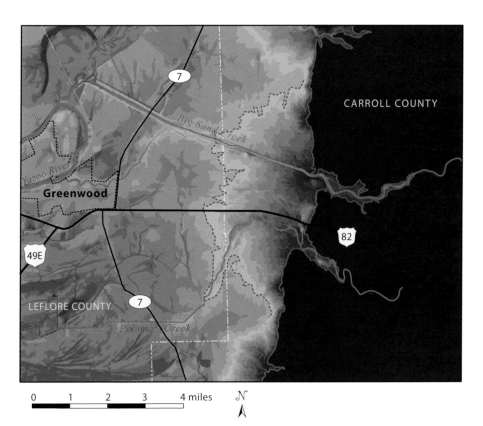

Digital elevation model of the alluvial fan east of Greenwood. The contour interval is 4 feet. The elevation of the alluvial fan in the outer margins of its apron (dashed line), to the west, is 128 feet above sea level, and the elevation of the bluff (solid red line) is 180 feet above sea level.

Looking west over the Delta from US 82, just east of Greenwood, at the apex of the large humid alluvial fan.

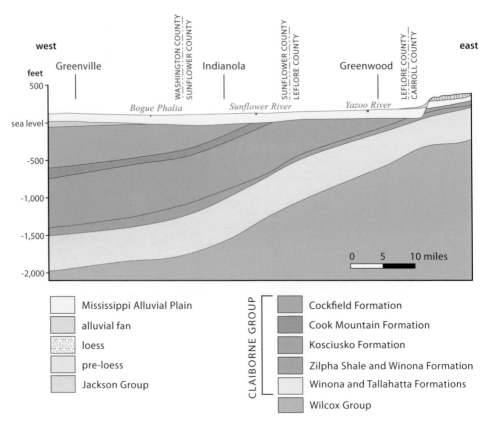

Cross section paralleling US 82 from just west of Greenville to just east of the bluff line. (Modified from Jennings 2001.)

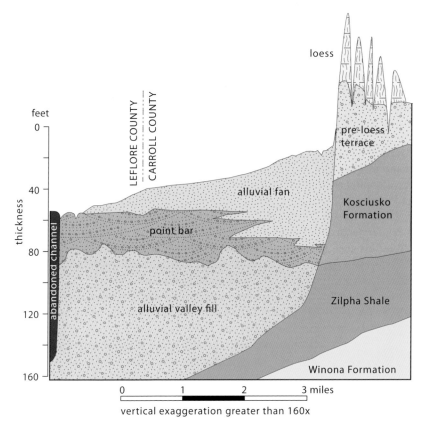

Cross section paralleling US 82 about 2 miles to the south, showing the relationship between the Mississippi River Alluvial Plain and Loess Hills physiographic provinces. (Modified from Kolb et al. 1968.)

which the Mississippi River Valley Alluvial Aquifer is recharged. Because fans occur all along the bluff line, they may play an important role in aquifer recharge. Groundwater percolating through the fans is thought to be a significant source of recharge water for the aquifer, and possibly other aquifers connected to it. The fan that parallels US 82 is well studied and documented by borings.

MISSISSIPPI 1 (GREAT RIVER ROAD)
Mayersville—Greenville—Rosedale
78 miles
See the map on page 138.

Between Mayersville and Rosedale, MS 1 travels mostly across sandy point-bar deposits of the western meander belt. It also crosses abandoned channels of the ancestral Mississippi and Ohio Rivers or abandoned courses of their crevasse streams and distributaries. Rivers and streams occupy some of the abandoned channels. Notable examples include Steele Bayou, about 6 miles north of Mayersville, and Lake Washington, west of MS 1 between the communities of Foote and Erwin.

From Mayersville to the junction with MS 14, MS 1 crosses the western meander belt, consisting of Holocene-age deposits that range up to 4,800 years old in this area. From the junction with MS 14 to Hampton, the road crosses Holocene-age backswamp deposits and is then back on the western meander belt to Wayside. Notable ridge and swale topography is visible east of MS 1 north of Yazoo Refuge Road in Foote. Ridge and swale topography develops as a low-gradient river migrates across its floodplain, leaving behind a series of point bars that creates the undulating terrain.

For 7 miles north of the MS 438 junction at Wayside, MS 1 travels on Pleistocene-age braided-stream deposits that are just under 15,000 years old. On the south side of Greenville, at Colorado Street, it crosses back onto point bar deposits, oxbows, and abandoned channels of the western meander belt.

The state boundaries along the Mississippi River were established in the 1800s. Though the political boundaries are fixed, the channel is not, which is why today's state boundary no longer perfectly parallels the river. There are several places along the river where natural and human processes have resulted in course diversions and meander cutoffs. For example, Lake Ferguson is an oxbow lake adjacent to Greenville, but it is not a natural oxbow lake. Following the flood of 1927, three

Winterville Mounds

About 6 miles north of the junction with US 82, MS 1 travels past Winterville Mounds, an official state historic site listed on the National Register of Historic Places. Named for a nearby community, the mounds are part of a prehistoric ceremonial center built by a Native American civilization that thrived from about AD 1000 to AD 1450. Twelve of the site's largest mounds, including the 55-foot-high Temple Mound, are currently the focus of a long-range preservation plan. The Winterville site is 2 miles north of the northern shore of Lake Ferguson and in the midst of several meander scars, which are visible on aerial photos. Considering the site was inhabited as recently as 600 years ago, it's likely that the Mississippi River was not any closer to the settlement during its occupation than Lake Ferguson is today.

The 1927 Mound Landing Levee Breach

A popular stop on many geology field trips is just west of the Great River Road. From Scott, take Ranch Road west 2.8 miles to the levee. Turn south and travel 1.8 miles to the Mound Landing crevasse site adjacent to the levee. On April 21, 1927, the Mississippi River breached the engineered levee here, causing the Delta to be flooded.

The breach at Mound Landing stands as one of the most destructive levee breaches in US history. Hundreds of men worked through the night to try and save the levee. In the end, they failed, and many were swept away as the floodwaters tore through the levee. The force of the water left its mark on the landscape, gouging out the depressions that are today occupied by Mound Crevasse and Grassy Lake. Once the floodwater had receded, the crevasse was a nearly 0.75-mile-long gap in the levee. The sediment composing the splay deposit, which the floodwaters left behind, is still visible in the fields east of the modern levee. One study suggests that as much as 20 percent of the floodwater related to the 1927 flood passed through the Mound Landing crevasse.

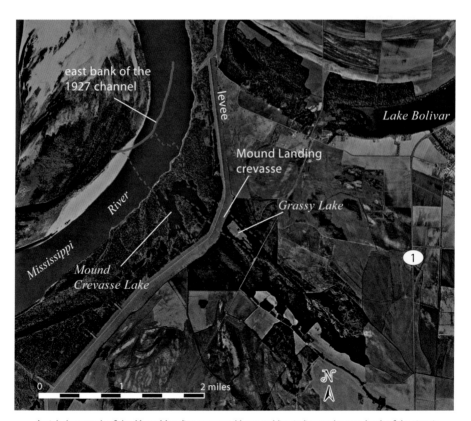

Aerial photograph of the Mound Landing crevasse. Heavy red line indicates the east bank of the river in 1927, and the dotted red lines indicate the extent of the breach during the flood.

large pronounced meanders north of Greenville, including Lake Ferguson, were purposefully cut off in order to straighten the river. This was one of the early actions taken by the Mississippi River Commission following the flood in order to decrease flood potential in the future. The river was shortened by 37 miles in order to enable water to pass through the area faster in the straightened channel. After digging trenches across the meander loops, explosives were detonated in the main channel on either end of the trenches, opening them up to the river. Lake Bolivar, west of MS 1 just north of Scott, fills an abandoned channel.

MISSISSIPPI 7
Greenwood—Grenada
30 miles
See the map on page 138.

From the US 82 junction at Greenwood to the CR 350 junction at Teoc Creek, MS 7 travels along the boundary between the eastern meander belt, with deposits ranging between 5,800 and 7,500 years old, and Pleistocene-age braided-stream deposits. From there it crosses braided-stream deposits for about 3 miles, until it begins to climb the alluvial fan created by Potacocowa Creek. About 1 mile north of Leflore, MS 7 crosses into the Loess Hills physiographic province. From here to Grenada the road travels along the southern side of the Yalobusha River on its floodplain and terrace alluvium. Along this stretch the road passes two loess-capped hills underlain by pre-loess alluvial sand and gravel and sand and clay of the Eocene-age Claiborne Group: at 1 mile and 6 miles east of the MS 35S junction at Holcomb.

The area between Greenwood and Avalon is interesting because of the variety of water bodies present in such close proximity. Sharkey Bayou, east of the road, is not an abandoned channel, as one might suspect; rather it is located within the braided-stream deposits laid down approximately 14,000 years ago, at the tail end of the last glaciation, by the ancestral Ohio and Mississippi Rivers. Since that time, neither river has flowed through this region. Potacocowa and Teoc Creeks flow from the bluff line and have laid down alluvial fans. Goose Pond, west of the road, is an oxbow lake, and the Yalobusha River occupies an abandoned channel of the ancestral rivers mentioned above.

MISSISSIPPI 16
Yazoo City—Canton
24 miles
See the map on page 138.

Between Yazoo City and Canton, MS 16 passes through the Loess Hills physiographic province. The hills are made up of windblown silt and underlying pre-loess alluvial sand and gravel. The silt was blown here from the west at the end of Pleistocene time. At Benton, the road crosses into the Jackson Prairie physiographic province, which is underlain by the Eocene-age Jackson Group, primarily the Yazoo Clay.

An exposure of loess is visible on the south side of MS 16 about 5.5 miles east of the US 49 junction at Yazoo City. Note the vertical nature of the outcrop's erosional face. The high permeability of loess enables water to travel through it vertically, so there is little runoff to erode exposed surfaces. However, loess does weaken, and chunks of it do cleave off, like glaciers do when they reach a

Kudzu

Kudzu smothering trees as it takes control of the landscape in the Delta.

No travel guide through the Delta would be complete without some mention of kudzu, because it's everywhere in the Delta. Kudzu (*Pueraria lobata*), native to southeastern Asia, was brought to the South in the 1880s. Southerners rapidly adopted the plant and used it in ornamental plantings, to provide shade from the intense summer sun, and as a protein supplement in cattle feed. With growth rates of up to 1 foot per day, kudzu was initially declared a wonder plant. In the 1930s and 40s the federal government recommended its use as a soil and slope stabilizer. By 1970 the plant was listed by the United States Department of Agriculture as a common weed. And in 1997 the plant, informally known as the "vine that ate the South," was finally declared a noxious weed. Noxious weeds are injurious to public health, recreation, wildlife, property, or agriculture. They are invasives that outcompete all indigenous species in an area.

body of water. You can see that it is weakly consolidated and very porous and permeable. The vertical fractures are related to tensional stress acting on the exposed face; root growth, which tends to be vertical; and the presence of calcite. The weathering and dissolution of calcite is thought to weaken the loess and promote failure. The loess veneer along this part of MS 16 is no more than a few feet thick, and it thins from here and is not present near the I-55 junction. Where the loess has eroded away, the road crosses the Yazoo Clay.

About 0.5 mile after the turnoff to Deasonville, MS 16 begins to descend toward the Big Black River on its Quaternary-age floodplain and terrace alluvium. The road crosses these deposits for just less than 2 miles.

INTERSTATE 20
VICKSBURG—BOLTON—JACKSON
44 miles

I-20 crosses the Louisiana-Mississippi state line on the bridge above the Mississippi River. At low flow Oligocene-age limestone of the Vicksburg Group is exposed at the base of the bluffs on the Mississippi side of the river. The Vicksburg Group was deposited in a shallow tropical sea that inundated the Mississippi Embayment. The rocks record a complete cycle of transgression and regression. The shallow-marine limestone and marl of the Mint Spring,

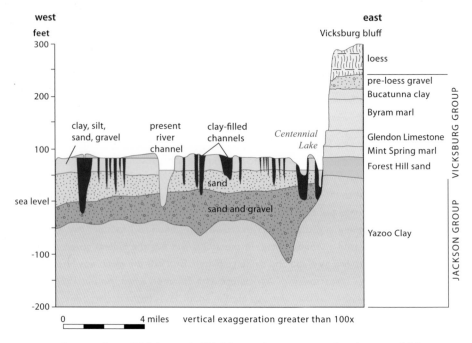

Cross section at Vicksburg. Modified from US Army Corps of Engineers, n.d.(a).

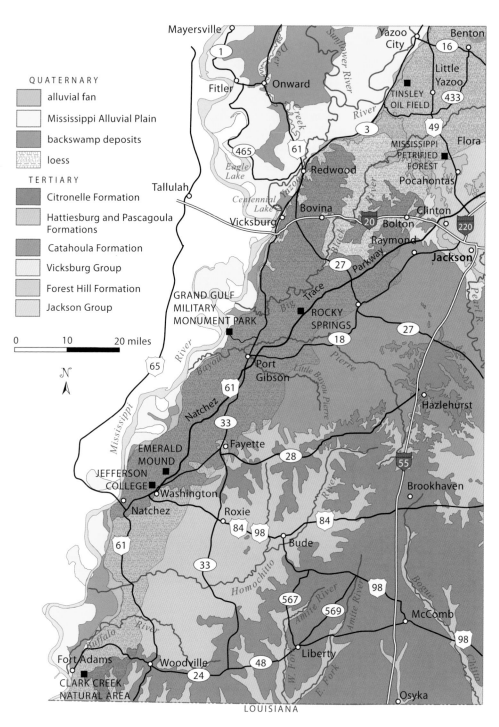

QUATERNARY

alluvial fan

Mississippi Alluvial Plain

backswamp deposits

loess

TERTIARY

Citronelle Formation

Hattiesburg and Pascagoula Formations

Catahoula Formation

Vicksburg Group

Forest Hill Formation

Jackson Group

0   10   20 miles

N

Mayersville
1
Fitler
Onward
Yazoo City
Benton
16
Little Yazoo
433
TINSLEY OIL FIELD
3
49
Flora
MISSISSIPPI PETRIFIED FOREST
465
61
Redwood
Pocahontas
Tallulah
Eagle Lake
Centennial Lake
Bovina
Clinton
Vicksburg
20
Bolton
220
Raymond
Jackson
27
Parkway
Big
Trace
ROCKY SPRINGS
27
GRAND GULF MILITARY MONUMENT PARK
18
Pearl R.
Port Gibson
Little Bayou Pierre
Pierre
61
Natchez
33
Hazlehurst
Fayette
28
River
55
EMERALD MOUND
JEFFERSON COLLEGE
Washington
Brookhaven
Natchez
Roxie
84
98
98
84
Bude
River
61
33
84
Homochitto
567
569
98
McComb
98
Fort Adams
Buffalo River
Woodville
48
Liberty
Amite River
W. Fork
E. Fork
Bogue
Chitto R.
24
CLARK CREEK NATURAL AREA
Osyka
LOUISIANA
65

Geology along I-20 between Vicksburg and Jackson.

Vicksburg

The City of Vicksburg is located on the bluff line marking the boundary between the Mississippi River Alluvial Plain and Loess Hills physiographic provinces. The loess in the Vicksburg area can be over 100 feet thick and was deposited by wind between 25,000 and 16,000 years ago. The loess forms an almost vertical wall when eroded, sometimes resulting in unstable slopes. You can see "calving" loess along any road in the area. Vertical fractures promote calving, and there are several reasons they form: tensional stress along the face of the weakly cemented loess tends to pull blocks away from the face of exposures, forming and widening fractures, but roots penetrating the loess and the groundwater-dissolution of calcite in the loess also help weaken the face of exposures.

The US Army Corps of Engineers Waterways Experiment Station at Vicksburg (now called the Engineer Research and Development Center) was the home of a 200-acre scale model of the Mississippi River Basin. In the 1940s, due to the wartime effort and a shortage of workers, Italian and German prisoners of war were used to sculpt the scale model, which represented the entire Mississippi watershed, including the Rocky and Appalachian Mountains. It was one of the most complex hydraulic models ever built. To this day the corps relies on scale models to understand how natural systems operate and how best to mitigate natural disasters.

Retaining walls against the loess bluffs at the Ameristar Casino parking lot, along the bluffs just north of the I-20 bridge. Inset: Steel braces support soil nails that anchor the retaining walls to the bluff.

The unstable loess bluffs have been a formidable obstacle to the development of Vicksburg's waterfront. Between 1930 and 2004, piers for the US 80 bridge moved nearly 2 feet as a result of movement along the bluffs, and the eastern bridge piers of I-20 moved up to 0.75 foot. The I-20 movement resulted in bridge work in 2004. The most recent movement was attributed to the development of a parking lot for the Ameristar Casino at the base of the bluffs.

Vicksburg has experienced notable flooding. The flood of 2011 set new record stages at Vicksburg. The peak stream flow of 2,310,000 cubic feet per second exceeded both the estimated peak of the Great Mississippi Flood of 1927 (2,278,000 cubic feet per second) and the measured peak of the 1937 flood (2,080,000 cubic feet per second).

Historic high-water marks on the flood wall at the west end of Clay Street in Vicksburg. The Mississippi in this photo is at 23 feet, 34.1 feet lower than the 2011 flood-crest record marked on the wall.

The Lower Mississippi River Museum of the US Army Corps of Engineers is dedicated to the lower portion of the Mississippi River. It is located at 910 Washington Street, near the flood wall at the western terminus of Clay Street. The museum has an outdoor scale model of the Lower Mississippi River and features displays and models showing river flooding as well as levee construction.

Vicksburg National Military Park

Most of the Civil War battle and siege of Vicksburg was fought on loess—across the deep ravines eroded in it, within earthen forts constructed of it, and in trenches and tunnels dug in it. Vicksburg residents dug "caves" into loess during the siege to escape cannon fire, and the high loess bluffs at Vicksburg overlooked a bend in the Mississippi River, thus providing a strategic location from which to control transportation along the river.

The entrance to the park is north of I-20 at exit 4 on Clay Street. The park contains monuments to troops who were involved in the 1863 battle and siege. They are constructed of various types of stone, some native to the state that the monument

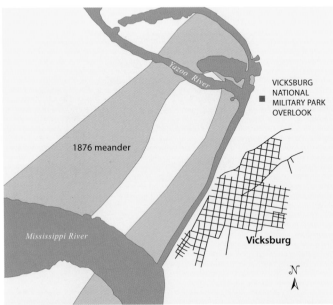

Map of the current (dark blue) and historic (light blue) Mississippi River at Vicksburg. The meander was cut off in 1876 and is now occupied by the Yazoo River.

Looking west from the loess bluffs in the Vicksburg National Military Park. The Yazoo River currently occupies the Civil War–time meander of the Mississippi River.

View into back of a redoubt. Although the terrain was formidable, the loess was easily excavated and shaped using hand tools. Erosion over the past 150 years has taken a toll on the loess earthworks in the park, therefore many features have been obscured.

View from a tunnel through the hill known as Thayer's Approach. The Confederate troops occupied the top of the hill. The valley in the center of the photo is where the Union forces started excavating a 6-foot-deep trench.

honors. Historical redoubts and redans, constructed of loess, are preserved in the park. A redoubt is a square refuge generally constructed of earthen material and enclosed on all sides, whereas a redan is triangular and open to the rear. Many of the trenches excavated in loess during the war are also still visible throughout the park.

In one of the more unique examples of wartime geoengineering, Union troops dug a trench up a hill toward a Confederate fort because the rugged terrain of the loess proved too formidable for direct assault. This campaign was called Thayer's Approach, named for the Union general commanding the operation. The plan was to create a cavity in the loess beneath the fort and destroy it with black powder. The war ended before the plan could be fully executed. The site of this historic event is preserved in the northern part of the park.

Marianna, and Glendon formations record the sea level rise, which was followed by the deposition of the Byram Formation sand and clay and Bucatunna Formation clay as the sea shallowed and the coastline grew seaward. The Marianna and Glendon Limestones represent the last major marine transgression of the sea into the Mississippi Embayment.

After leaving the bridge the interstate enters the Loess Hills physiographic province. The river bluffs and hills are made up of windblown silt capping pre-loess alluvial gravel, all of Pleistocene age. Underlying these deposits are the limestones and marls of the Vicksburg Group and clay, sand, and sandstone of the Catahoula Formation. Although loess was deposited farther east, the best exposures on I-20 are between the river and Bovina. Vertical cliffs stand in relief on both sides of the highway.

As you travel eastward on I-20 the loess thins, but several good exposures are visible between Vicksburg and the Big Black River. Most of the loess exposures contain terrestrial gastropod shells and calcite concretions, informally called

A loess exposure on I-20 between eastbound mile markers 6 and 7. Vertical fracturing, calving, and terrestrial gastropod shells can be observed.

"loess dolls," and fractured and calving loess blocks. The concretions precipitate lower in the loess section from percolating groundwater, which dissolves calcium carbonate–rich minerals higher in the bluff.

East of Bovina, between exit 15 and just west of mile marker 18, I-20 crosses the Big Black River and its associated floodplain. Research is ongoing to determine the exact origin of a 3-mile-long, anomalously straight natural section of river. As I-20 nears exit 27 at Bolton, the loess capping the Vicksburg Group limestones becomes thin, and it is very thin to nonexistent east of Bolton.

Between mile marker 31 and exit 36, I-20 travels through the Southern Pine Hills physiographic province, predominately underlain by Catahoula Formation sand and clay, Vicksburg Group limestone, and Forest Hill Formation sand and clay of Oligocene age. Between exits 33 and 34, the increase in juniper and cedar trees correlates with the underlying Vicksburg limestone; the trees thrive in the alkaline soil derived from it, whereas other species cannot.

From exit 36 to its junction with I-55 in Jackson, I-20 travels through the Jackson Prairie physiographic province, which is underlain by the Eocene-age Yazoo Clay of the Jackson Group.

US 49
JACKSON—FLORA—YAZOO CITY
36 miles
See the map on page 155.

From its junction with I-220 at Jackson to near Pocahontas, a distance of about 9.5 miles, US 49 travels through the Jackson Prairie physiographic province, which is underlain by the Yazoo Clay of the Eocene-age Jackson Group. The Yazoo is well known for its shrink-swell properties. It expands when wet and shrinks when dry, resulting in uneven roadways and shifting foundations. The Yazoo does have commercial value. Weathered clay loses its shrink-swell potential, and a local manufacturer used it for many years to make bricks. The gently rolling terrain is produced as the clay erodes to a relatively flat plain. There are, however, few if any exposures. From about Pocahontas to just south of Flora, US 49 travels through sand of the Oligocene-age Forest Hill Formation.

From just south of Flora to Yazoo City, US 49 travels through the Loess Hills physiographic province. The hills are made up of windblown silt capping pre-loess alluvial gravel, all of Pleistocene age. Below the pre-loess gravel lies the Jackson Group, primarily the Yazoo Clay. At the Madison-Yazoo county line, US 49 crosses the Big Black River and its associated floodplain and terrace alluvium. The alluvium spans from about 1 mile south of the river to about 2 miles north of it. Numerous outcrops of loess are visible north of Little Yazoo along this very hilly section of US 49. Some are natural exposures on hilltops, such as the one on the east side of the road about 0.5 mile north of Castle Chapel Road, and another on the west side of the road north of the Yazoo County High

Mississippi Petrified Forest

In Flora, just west of US 49 off Petrified Forest Road, there is geologic locale well worth a side trip. The Mississippi Petrified Forest, a National Natural Landmark, is situated on private land that is open to the public. The landmark contains an earth science museum and nature trails, along which numerous large petrified logs have weathered out of sand of the Forest Hill Formation. The entire forest lies within this Oligocene-age sand, which was deposited on a floodplain near a delta. At least two trees are reported to be in growth position, but it is believed that the rest were transported by floodwater and subsequently buried. The original cellulose in the trees was fully replaced by silicon dioxide in a process known as *petrification*. The cell-by-cell replacement process was so complete that botanists have been able to identify several species, such as fir, maple, sequoia, and an extinct conifer called *Cupressinoxylon florense*, named after the town of Flora.

In the early 1900s, conservation efforts were nonexistent and soil erosion was a common problem in the Loess Hills province. Up until the early 1960s, when the current owners took control, this area was left to erode naturally. The heavily forested area today in no way resembles the heavily eroded and barren landscape of the 1930s. Although the erosion of loess and the Forest Hill sand is how the petrified logs were discovered, the site would not have been preserved for future generations without conservation efforts.

Petrified logs in the Mississippi Petrified Forest.

Tinsley Oil Field

In Little Yazoo, at the junction with Dover Road, a roadside historical marker in the northbound lane notes the Tinsley Oil Field, which is actually 4 miles west of US 49. It was the first commercial oil field in Mississippi. Surface geologic mapping performed by Frederic F. Mellen of the Mississippi Geological Survey, with funding from the Works Progress Administration, identified a geologic structure, which was successfully drilled in 1939. The initial production was from the Woodruff Sand in the Selma Chalk. Since that time numerous other Cretaceous-age reservoirs have been productive, including ones in the Eutaw and upper Tuscaloosa Groups and the Paluxy, Mooringsport, Rodessa, Pine Island, and Hosston Formations. Production has also been established in the younger Wilcox and Claiborne Groups (Eocene age) and the older Smackover Formation (Jurassic age). The Tinsley Oil Field is Mississippi's largest oil field and has produced more than 240 million barrels to date.

G. C. Woodruff No. 1, the Tinsley Oil Field's discovery well, was the first to produce oil south of the Ohio River and east of the Mississippi River.
—Courtesy of Frank Noone

The rolling topography along US 49 between Little Yazoo and Yazoo City is attributed to the increased thickness of the loess, proximity to the bluff line, and the westward direction of drainage.

School. Other exposures can be seen in sand and gravel pits that are mining the pre-loess deposits.

The change in topography north of Little Yazoo is related to several factors. First, as US 49 approaches Yazoo City and the bluff line, the thickness of the loess increases. The increased thickness provides for more relief when it is eroded. Second, because the bluff line is near, with its relatively large change in elevation, streams are more actively eroding the loess. And third, there is a change in drainage direction north of Little Yazoo. South of here the streams drain south to the Big Black River, whereas north of Little Yazoo they drain west to the Yazoo River. US 49 crosses the stream valleys perpendicular to them, creating the road's up-and-down feel.

US 61
Louisiana—Natchez—Vicksburg—Onward
145 miles
See the map on page 155.

From the Louisiana border to Redwood, US 61 travels through and on the Loess Hills physiographic province. The hills are made up of windblown loess that caps pre-loess alluvial sand and gravel, all of Pleistocene age; several older formations, of Miocene and Oligocene age, lie below the pre-loess sand and gravel.

From the Louisiana border to 2 miles north of MS 24, US 61 crosses a thin veneer of Citronelle Formation sand. North of MS 24 clay and sand of the Hattiesburg and Pascagoula Formations are exposed. From Louisiana to about 17 miles north of the MS 24 junction at Woodville, the loess is relatively thin and sporadic. The thickness of the loess decreases rapidly to the east as well. US 61 also crosses streams and their associated floodplain and terrace alluvium, including the deposits of the Buffalo River, about 10 miles north of MS 24, and the Homochitto River, about 19 miles north of the junction.

From its junction with US 84W at Natchez to about 9 miles north of the Big Black River, the sand and sandstones of the Catahoula Formation lie beneath the loess and pre-loess alluvial sand and gravel. This segment of US 61 again crosses streams and their associated floodplain and terrace alluvium, including Little Bayou Pierre and Bayou Pierre, near Port Gibson, and the Big Black River.

Clark Creek Natural Area

The Clark Creek Natural Area is a 700-acre tract located about 14 miles west of Woodville. It contains about fifty waterfalls, some up to 30 feet in height. The waterfalls are in tributaries of Clark Creek that have eroded down through the loess and pre-loess gravel to the Pascagoula Formation. The Pascagoula is primarily composed of clay, but in places it features relatively well-cemented, erosion-resistant siltstone, which is the ledge former from which the waterfalls cascade. The clay erodes much more easily and creates the scenario in which differential erosion leads to a waterfall.

A 30-foot waterfall in the Clark Creek Natural Area.

Rills that form on top of the more resistant siltstone in the Pascagoula Formation.

Boulders of the Pascagoula that washed out of the plunge pool during storms.

Upstream of the waterfalls, the tributaries flow over the cemented siltstone layer until they reach the falls. The water carries sand and gravel, which abrade narrow gullies (rills) in the silty clay face of the falls. As the Pascagoula is undercut, the ledge of the falls periodically fails and breaks off. As a result, the knickpoint, or location of the waterfall, advances farther upstream. Downstream of the waterfalls, large boulders of the Pascagoula eroded from beneath the falls lie in the channel. Their size attests to the power of runoff during storm events.

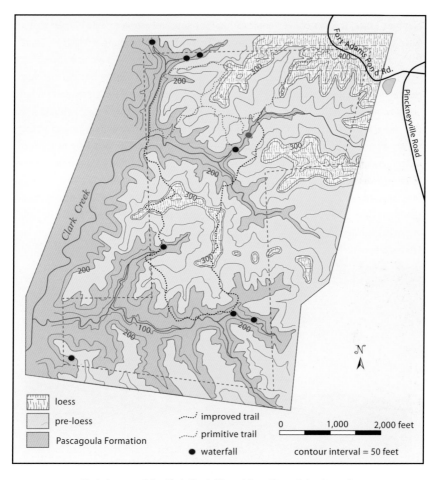

Geologic map of the Clark Creek Natural Area. The red dot shows the waterfall featured in the previous image.

Natchez

Approximately 2.5 miles west of US 61, the historic downtown portion of Natchez sits on top of a 200-foot bluff that overlooks the Mississippi River. The lower portion of the bluff consists of approximately 80 feet of pre-loess alluvial sand and gravel, and the upper portion consists of loess.

Natchez is the oldest city on the Mississippi River, having been established in 1716. It once served as the capital of the Mississippi Territory. The city occupies a strategic location on the bluff overlooking the Mississippi River. As the southern terminus of the Natchez Trace, it was a convenient location for pilots and crews to unload their cargo, leave the river, and head back north on the Trace. Today the modern Natchez Trace Parkway, which commemorates this route, still has its southern terminus in Natchez.

Though the bluffs at Natchez have always been prone to failure, this did not stop residents from building near the edge to take advantage of the view. Slides and slumps are a continual threat. The precariously perched Clifton Heights neighborhood is in the most desperate situation. Fifteen homes along Clifton Avenue were originally constructed 50 feet from the bluffs in the 1880s. In 1951 a 400-foot-long portion of Clifton Avenue was removed by a major slide in the loess. Some of the homes now have as little as 8 feet between the front door and the bluffs. The US Army Corps of Engineers and city engineers have studied the geology of the bluffs to determine if the homes can be saved. In 2006, in a report to the mayor of Natchez, geologist John Bornman warned, "Let us not ignore the warnings of the past. Construction along the bluff will eventually wind up under the hill."

Looking north along Clifton Avenue in downtown Natchez. The original roadway was destroyed by slope failure along the top of the bluff, and only a walkway is present today.

Jefferson College

Historic Jefferson College is located in Washington, 1 mile north of the US 84/98 junction. Washington was the territorial capital in 1811. The General Assembly of the Mississippi Territory authorized the college in 1802, and Governor William C. C. Claiborne stated that it would "become a fruitful nursery of science and virtue." It opened in 1811 as the first educational institution in the Mississippi Territory and included a museum of natural science as well as course work in mineralogy. One of its students, Benjamin L. C. Wailes, became a noted geologist and later taught at Jefferson College and served on its board of trustees. Wailes authored the first report on Mississippi's geology, which was featured in the *Report on the Agriculture and Geology of Mississippi*, published in 1854. Wailes also donated many mineral and fossil specimens to the college's museum.

Today Jefferson College is a historical site with a museum and nature trail. The museum has displays on Wailes, including his geological contributions. The location of the college was selected to take advantage of Ellicotts Springs. The original spring, which supplied water to the college, can still be seen along the nature trail. The trail also leads to Saint Catherine Creek, where exposures of loess and pre-loess terrace deposits can be seen. One can easily walk the creek during low-flow conditions in order to see exposures up close.

Grand Gulf Military Monument Park

North of Port Gibson, about 8 miles west of US 61 from the junction with MS 462, the Grand Gulf Military Monument Park displays in its museum arrowheads, petrified wood, minerals, rocks, and fossils from the area. One of the more notable specimens is a mastodon leg bone with tool marks indicating that this giant was brought down by Paleo-Indian hunters.

In the park, the Oligocene-Miocene-age Catahoula Formation sandstone is exposed in the bluff along the access road to Fort Cobun. Rivers and streams deposited the sand of this sandstone in a delta or on a coastal plain. The Catahoula is recognized as one of the major ledge-forming units in southern Mississippi. It's also known for its abundant petrified palm trees and very hard quartz- and opal-cemented layers. The well-cemented layers are locally called "quartzites," a term generally used to refer to metamorphosed sandstone, but Catahoula layers are not metamorphic in origin. The proper name is *orthoquartzite*. Like quartzite, these rocks do break across individual

There are several locations along Grand Gulf Road where loess has been overrun with kudzu, which has preserved near-vertical walls. Here the 30-foot-tall embankment stands at 65 degrees!

Looking north into the Fort Cobun ruins. The hillslope in the back right-hand side is the east-facing loess bluff. The 6-foot-tall embankment to the left is the wall of the fort.

grains because the cement is as strong as the grain; sandstones, however, typically break around grains because the cement is not as strong as the individual sand grains. Several localities in Claiborne County feature Catahoula cemented with opal, some of which is gem quality. The source of the silica for this opal is ash. It was blown into the region from volcanic activity in western North America during Oligocene and Miocene time.

Of the two earthen forts in the park, Fort Cobun is the most impressive, at least from a geological viewpoint. The fort commanded a strategic location on the cut bank at Rock Point, 70 feet above the Mississippi River. The Catahoula at water level inhibited the river from eroding into the bluff, instead diverting it westward. Today the river lies to the west of its position during the Civil War. Unlike most fortifications of this vintage, which troops constructed by building loess walls, Fort Cobun was trenched into place. Confederate troops removed the loess down to the top of the Catahoula and were protected behind a 40-foot-thick wall of loess. The fortification enabled them to prevent Union troops from landing and taking Port Gibson.

A 20-foot-tall exposure of Vicksburg Group limestone, on the east side of US 61, about 6 miles north of the I-20 junction. This is a good site for collecting bivalve (clam) and echinoid (sea urchin) fossils.

From about 9 miles north of the Big Black River to about 5 miles north of the I-20 junction in Vicksburg, limestone and marl of the Oligocene-age Vicksburg Group lie beneath the loess and pre-loess alluvial sand and gravel. About 5 miles north of the I-20 junction, US 61 drops down in elevation onto the Mississippi River Alluvial Plain. From here to the MS 3 junction at Redwood, US 61 travels along the base of the bluff line, to the east. Along this stretch the loess, pre-loess sand and gravel, and Vicksburg Group limestone are visible in the bluffs. The limestone was deposited in a shallow sea during the last major transgression into the Mississippi Embayment.

From its junction with MS 3 at Redwood to just north of the Issaquena-Sharkey county line, US 61 travels on point bar deposits of the Ohio and Mississippi Rivers, except where it crosses the Yazoo River and Deer Creek. These modern streams occupy abandoned courses of the ancestral rivers. The modern Mississippi is interpreted to have flowed through here between 2,600 and 6,200 years ago. The highway transitions onto backswamp deposits starting 4.5 miles north of the first Deer Creek crossing, just north of Redwood, and on into Onward. The clayey sediment indicates that the area was subjected to floodwater deposition during the Holocene, but it lies between the eastern and central meander belts and therefore does not feature any oxbows or channels from that period. However, Deer Creek flows through the center of the backswamp deposits and is likely flowing through an older abandoned channel that was present prior to the deposition of the backswamp clay. The Delta National Forest is situated on these backswamp deposits.

Delta National Forest

The Delta National Forest comprises nearly 61,000 acres in Issaquena and Sharkey Counties. Located on Holocene-age backswamp deposits of the Mississippi River Alluvial Plain physiographic province, the forest harbors flora and fauna that have thrived in the Delta for thousands of years.

For most of the past 10,000 years the Delta has been constantly changing due to natural causes: river channels migrated, meanders were cut off, and channels were abandoned. The technological advances of the twentieth century, however, have changed that. The only oxbows that have formed on the Mississippi River since 1927 were artificially created and engineered to reduce flooding or straighten the Mississippi's channel for navigational purposes. The creation of levees and the confinement of all rivers in the province, not just the Mississippi, have altered the natural processes that ruled here for thousands of years. Soil nourishment, once accomplished by flooding rivers, only occurs when levees fail or back flooding takes place. (Back flooding occurs when tributary streams cannot flow into the Mississippi because the Mississippi is too high.)

The US Army Corps of Engineers built four greentree reservoirs in the 1980s to mitigate wetland loss in the national forest. These managed areas help ensure that native plants and animals and migratory birds experience flooding regimes similar to the conditions in which they evolved. The reservoirs are typically flooded in the late fall and winter. They must be drained before the trees start growing in the spring.

US 84

Natchez—I-55
62 miles
See the map on page 155.

US 84 crosses the Louisiana-Mississippi state line on the bridge above the Mississippi River at Natchez. The bluff that formed on the Mississippi side of the river, part of the Loess Hills physiographic province, stands 200 feet above the river. They are composed of loess that caps pre-loess alluvial sand and gravel, including the Natchez Formation; all are of Pleistocene age. The Natchez Formation is 70 feet thick in the bluff and was formed as braided streams, sourced by glacial meltwater, flowed over the alluvial plain.

Between the US 61 junction, at Natchez, and Bude, US 84 and US 98 are joined. From the junction with US 61, US 84/98 travels through loess hills and the underlying gravel, both of which thin toward the east.

Near the Adams-Franklin county line the highway passes into the Southern Pine Hills physiographic province. From this boundary to about 3 miles east of the Franklin-Lincoln county line, US 84/98 (US 98 breaks off near Bude) travels on the Hattiesburg and Pascagoula Formations, which consist of clay, sandy clay, siltstone, and sand. The contact between these formations is gradational and difficult to determine; the nonmarine sediment of the Hattiesburg is marked by fossil leaves, and the estuarine sediment of the Pascagoula is marked by bivalve fossils. Distinguishing these two formations when mapping south Mississippi is problematic, and doing so remains a priority for state geologists. On most state geology maps the formations are mapped as one unit as the Hattiesburg and Pascagoula Formations and have not been differentiated. From east of the Franklin-Lincoln county line to I-55, the high hills around US 84 contain gravel of the Pliocene-age Citronelle Formation.

US 84 also crosses the terrace and floodplain alluvium of modern streams. The most notable of these deposits belongs to the Middle Fork of the Homochitto River, about 10 miles east of the MS 33 junction. A leg bone of the Miocene-age rhinoceros *Teleoceras* was discovered in this creek bed.

An approximately 10-foot-tall exposure within the Hattiesburg and Pascagoula outcrop belt along US 84, about 1.6 miles east of the MS 184 junction.

US 98
Natchez—I-55
58 miles
See the map on page 155.

Between Natchez and Bude, US 98 and US 84 are joined. See the preceding road log for information about that stretch of highway.

For about 3 miles south of its junction with US 84 near Bude, US 98 travels on clay, sandy clay, siltstone, and sand of the Miocene-age Hattiesburg and Pascagoula Formations. The road then transitions onto sand and gravel of the Pliocene-age Citronelle Formation. It also crosses Quaternary-age terrace and floodplain alluvium of modern streams, notably that of the Homochitto River, just south of Bude, and the East Fork of the Amite River, near the MS 569 junction.

MISSISSIPPI 1 (GREAT RIVER ROAD)
Onward—Mayersville
26 miles
See the map on page 155.

At the junction of US 61, MS 1 is on fine-grained backswamp deposits, but it immediately passes onto point bar deposits, over which Deer Creek flows. From here to Mayersville the road travels mostly across sandy point-bar deposits, except for where it crosses abandoned channels and courses of the ancestral Mississippi and Ohio Rivers or those of their crevasse streams and distributaries. For example, Steele Bayou, between the Sharkey-Issaquena county line and the MS 465 turnoff, occupies an abandoned channel.

Between MS 465 and the community of Fitler, there is notable ridge and swale topography on the east side of MS 1 for about 1 mile. Ridge and swale topography develops as a river with a low gradient widens its valley, leaving behind a series of point bars that creates undulating terrain.

Between Fitler and Tallula, MS 1 parallels the engineered levee of the Mississippi River, which is on the west side of the road.

MISSISSIPPI 3
Redwood—Yazoo City
37 miles
See the map on page 155.

Between Redwood and Yazoo City, MS 3 travels on the Mississippi River Alluvial Plain, paralleling the bluff line, which is the boundary of the Loess Hills physiographic province to the east. The hills adjacent to MS 3 are made up of windblown silt that caps pre-loess alluvial sand and gravel, all of Pleistocene age. Below lie the Oligocene-age Vicksburg Group and Eocene-age Jackson Group.

The highway crosses several humid alluvial fans at the base of the loess hills; these develop where streams exit the hills and drop much of their sediment load onto the alluvial plain. These fan-shaped sand, and sometimes gravel, deposits form along the length of the bluff line, from Vicksburg to Tennessee. They are noticeable as slight rises in the road. The stream channels are near the apex, or top, of the fans.

About 1 mile north of the US 61 junction at Redwood, there is an abandoned course of the ancestral Ohio and Mississippi Rivers north of the road. The Yazoo River occupied it for a while until an artificial channel for the river was engineered farther west. The abandoned course still retains water. About 4 miles north of the US 61 junction, limestone of the Oligocene-age Vicksburg Group is exposed beneath the loess on the east side of MS 3.

NATCHEZ TRACE PARKWAY
NATCHEZ—RIDGELAND
101 miles
See the map on page 155.

The Natchez Trace Parkway is a National Park road that closely follows the course of the Trace, originally a Native American trail that ran from Natchez to Nashville. The French were the first to map it, in 1733. The Trace became an important wilderness road, the most heavily traveled in the Old Southwest, until steamboats provided a faster and safer mode of travel in the early 1800s.

Between its terminus at Natchez and mile marker 77, the Trace travels through the Loess Hills physiographic province. The hills are made up of Pleistocene-age windblown silt on top of the Oligocene-Miocene-age Catahoula Formation. The Catahoula consists primarily of sand, clay, and sandstone deposited by streams along a coastline that extended from Warren County

Emerald Mound

Just north of mile marker 10 is the National Historic Landmark known as Emerald Mound. The mound is certainly of archaeological interest, but it also represents a significant earthmoving project. Although Emerald Mound appears to be built on a natural hill, it is actually the second-largest prehistoric mound in the United States, second only to Monks Mound in Cahokia Mounds State Historic Site, in Illinois. Ancestors of the Natchez Indians built the structure; it covers nearly 8 acres and rises 35 feet above the surrounding terrain. It is believed they started building the mound around 1250, completing it around 1600.

eastward to Rankin County, and then down through Wayne County. Occasionally, Quaternary-age gravel outcrops between the loess and the Catahoula. These pre-loess terrace deposits are composed of sand and gravel that were laid down by streams that flowed through the area prior to loess deposition in the Pleistocene. The Catahoula is a ledge former and easily recognized where exposed. In south-central Mississippi, intervals within the Catahoula are well cemented. In areas where these sandstone units are exposed in outcrop, they stand out in relief as ledges because the silica cement is not easily weathered. Between mile markers 12 and 13 there is an excellent 90-foot-tall loess exposure, with pre-loess gravel at its base, in a cut bank along Turpin Creek.

Between mile markers 13 and 58 the Trace crosses several streams and their associated floodplain and terrace alluvium. The chocolate-brown and coffee-colored waters of Mississippi's streams and rivers usually startle visitors from western states. The coloring is caused by a large suspended load of clay and silt. However, at low flow the streams between mile markers 13 and 58 usually run clear, a pleasant surprise. The sand and gravel in the pre-loess terrace deposits, which form the banks and channels in many of the streams, contain less clay than other formations in the state. Most of the point bars in this area contain gravel weathered out of the terrace deposits that rest just below the loess. At Bayou Pierre, north of Port Gibson, there is an excellent example of point bar and cut bank development, just southeast of the bridge crossing. The river is simultaneously cutting the outer bank of a meander while depositing sediment in a point bar on the inside bank. This is how low-gradient streams and rivers migrate in the Delta.

View of the Bayou Pierre cut bank and point bar in the meander, as seen from the Natchez Trace bridge.

Rock hounds like the stream deposits that contain high-terrace gravels because they contain exotic rocks (at least by Mississippi standards), including volcanic welded tuffs and rhyolites from Missouri, the Sioux Quartzite from Minnesota, and Keokuk geodes from Iowa. Much of the gravel is chert containing Paleozoic-age marine invertebrate fossils that have been replaced by silica. Glaciers advancing from the north during the Pleistocene plucked these rocks from their outcrops and moved them south, where the ancestral Missouri, Ohio, and Mississippi Rivers later picked them up and redeposited them in Mississippi.

The Catahoula Formation is exposed where erosion has cut down through the loess. The eponymous spring at Rocky Springs was named for the outcrop of Catahoula sandstone it developed in. Because the Catahoula sandstone is resistant to erosion, waterfalls can develop in it. Waterfalls develop at knickpoints, where water flows over a resistant lip of rock and cuts into the softer rock below the resistant layer. When the overhang becomes too large to support itself, it collapses, and the knickpoint migrates upstream.

Natural springs were an important source of economic development in the state in the early 1900s. Several mineral spring resorts opened between Raymond and Jackson. One of the most famous was Cooper's Well, which

Owens Creek Falls flowing over Catahoula Formation sandstone at mile marker 52.4. Erosion has limited access to the trail, but the falls are still easily viewed from the parking lot.

Rocky Springs

Rocky Springs is located at mile marker 54.8. In the 1790s settlers were attracted to this area's rich soil and numerous springs. Springs form when groundwater percolating downward reaches an impermeable layer. Instead of continuing on to deeper horizons, the water is forced to move laterally; oftentimes it emerges at the surface. There are numerous layers in the Catahoula Formation of this region that are relatively impermeable because the pores have been filled with cement. By 1860 Rocky Springs was a community of more than 2,600. Between 1860 and 1920 the area was devastated by the Civil War, yellow fever, the destruction of cotton crops by boll weevils, and erosion. The spring for which the town was named has dried up, and today only a church with a cemetery still stands.

featured stinking water from the Catahoula Formation, 100 feet below the surface. Reverend Preston Cooper successfully promoted the well water as having medicinal properties, and the resort prospered from 1841 until the Civil War, when Union troops destroyed the site. It was later rebuilt and remained open until the 1920s. At one time the resort had a hotel that accommodated eight hundred guests. Unfortunately none of the ruins remain. Hydrogen sulfide gas was responsible for the rotten-egg smell. The gas forms when bacteria convert sulfate in the water to sulfide. Hydrogen sulfide does not pose a health risk and is common in groundwater systems.

Between mile markers 77 and 89, the Trace travels through the Pine Hills physiographic province. Between mile markers 77 and 83, the road travels through Catahoula Formation sands; between mile markers 83 and the junction with I-20, it's limestone of the Oligocene-age Vicksburg Group; and from I-20 to mile marker 89, it's the Oligocene-age Forest Hill Formation sand.

Between mile markers 89 and the I-55 junction at Ridgeland, the Trace travels through the Jackson Prairie physiographic province, which is predominantly underlain by Yazoo Clay of the Eocene-age Jackson Group, except between mile markers 96 and 98, where it again travels through Forest Hill Formation sand. The Yazoo Clay is exposed beneath the Forest Hill sand where the Trace drops to lower elevations.

SOUTHEASTERN MISSISSIPPI

Southeastern Mississippi is predominantly within the Southern Pine Hills phys-iographic province, but the northeastern corner falls within the Jackson Prairie and North Central Hills physiographic provinces, which are underlain by sand, clay, and limestone of the Vicksburg, Jackson, Claiborne, and Wilcox Groups. These formations (Oligocene through Eocene age) were deposited in and along a fluctuating shoreline as sea level rose and fell multiple times in the Mississippi Embayment. The Vicksburg sediments were deposited during the last major transgression in the embayment. Clay, sand, and gravel of the Catahoula, Hat-tiesburg, Pascagoula, and Citronelle Formations underlie the Southern Pine Hills province. The Catahoula, Hattiesburg, and Pascagoula Formations were deposited on a broad coastal plain and in deltas between Oligocene and Mio-cene time. The clay and silt of the Pascagoula represents a minor transgression in the embayment. The Pliocene-age Citronelle dominates southern Missis-sippi. Rivers scouring northern and central Mississippi sourced its sediments, which were deposited on a broad coastal plain in a range of environments, from braided river channels to deltas along the coast. The Gulf of Mexico receded to about where it is today as this formation developed.

The topography of southeastern Mississippi ranges from rolling hills to ridges and valleys dissected by dendritic drainage in the Mississippi, Pascagoula, and Pearl River watersheds. Dendritic drainage patterns resemble the branches of a tree: the main stream is the trunk and the tributaries are the branches. This type of drainage forms when bedrock is fairly uniform and not complicated by geologic structures, such as faults and folds, which can change the directions of flow. The formations of southern Mississippi generally dip toward the Gulf, but the angle of dip is very slight. Elevations in the province range from 580 feet near Meridian and Magee to approximately 50 feet in the Pascagoula River floodplain, in the southeastern corner of the state near Lucedale.

SALT DOMES

During the opening of the Gulf of Mexico, approximately 175 million years ago, as Pangaea rifted apart, a significant amount of salt was deposited in the Gulf, from Florida to Texas. Supersaturated seawater precipitated the Louann Salt (halite) in the Mississippi Interior Salt Basin; this salt formation exceeds 6,000 feet of thickness in places. The halite of the Louann Salt has a specific gravity of approximately 2.1, which means it is just over twice the weight of water per unit volume. The sediments deposited on top of the Louann have specific gravity values that are generally greater than that of halite. Quartz and calcite, two common components of the overlying sandstone and limestone, have specific gravity values of 2.65 and 2.70 respectively. The lower specific gravity of the salt and the pressure of the overlying sediment cause it to behave like soft modeling clay, becoming mobile as it is squeezed from its mother bed.

Often the easiest path for mobilized salt is vertical. The salt can move as a ridge or an inverted teardrop shape. Salt ridges may not rise very far, but they can deform surrounding sediment forming what are called salt anticlines or faulted salt anticlines. In addition to bending and faulting surrounding rocks, invading salt can even pierce and pass through formations as it rises toward the surface. These are called *piercements.* Salt rises in a fashion similar to hot wax in a lava lamp, stopping at a depth where the density difference between it and the

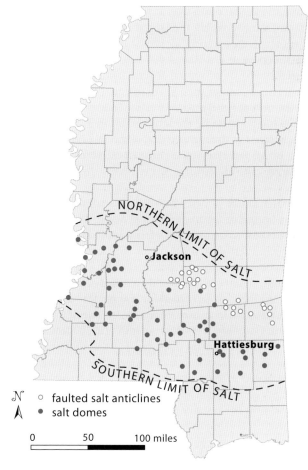

Salt domes and faulted salt anticlines within Mississippi's portion of the Mississippi Interior Salt Basin. (Modified after Thompson 2009.)

surrounding sediment is not great enough for further movement. The term *salt dome* comes from the domed shape of the rocks directly above the body of salt that moved. It is generally thought that salt movement has stopped in Mississippi, but in 2007 the Mississippi Office of Geology documented that the Ruth Salt Dome, in Lincoln County, moved as recently as the Quaternary Period.

The Mississippi Interior Salt Basin extends across south-central Mississippi and slightly into both Louisiana and Alabama. This region of Mississippi features fifty-three piercement domes and at least twenty-six faulted salt anticlines. Rock formations that are faulted, domed, or pierced by rising salt become potential traps for migrating hydrocarbons. Oil and natural gas flow through permeable rock formations, but they don't flow through salt. As they migrate horizontally through a formation, they become trapped where they encounter salt. Salt is the trapping mechanism for many of Mississippi's oil and gas fields.

Mississippi's most famous salt domes are the Tatum Salt Dome, in Lamar County, and the Richton Salt Dome, in Perry County. Both have histories related to our nation's entry into the atomic age.

The Richton Dome, possibly the largest and shallowest piercement dome in the state, encompasses an area of 7 square miles. Salt of this dome has been encountered at depths of less than 1,000 feet from the surface. The Richton Dome has been the focus of several federal investigations. In the late 1970s and early 1980s it was one of the US Department of Energy's top five potential storage sites for high-level nuclear waste. More recently, the Department of Energy selected the dome to be the latest addition to the US Strategic Petroleum Reserve. Because of their impermeability to oil, hollowed-out salt domes make excellent oil storage reservoirs. The plan called for the creation of sixteen ten-million-barrel caverns, but it was eventually dropped from the department's budget.

The Tatum Salt Dome is best known for its contributions to the United States' weapons-testing program. Following World War II the Atomic Energy Commission and the Department of Defense wanted to be able to monitor the underground nuclear testing of other nations but were unsure what the seismic signature of a subterranean blast would look like. The salt was likely chosen because of its ability to flow, which would eventually close the cavern created by the blast. Project Dribble, which called for two nuclear detonations within the Tatum Salt Dome, sought to answer this question. The first detonation, in solid salt, took place on October 22, 1964, at a depth of 2,700 feet. The blast opened a cavity approximately 110 feet in diameter. The second test, performed in the cavity created by the first test, occurred on December 3, 1966. Two additional bombs were detonated in 1969 and 1970, but they were not nuclear tests; the bombs were mixtures of oxygen and methane.

All that remains at the test site is a stone monument with a brass plaque that warns against drilling in the area. Locals feared that groundwater might be contaminated, so in 2000 the federal government built a water pipeline in the area to provide potable water from an outside source.

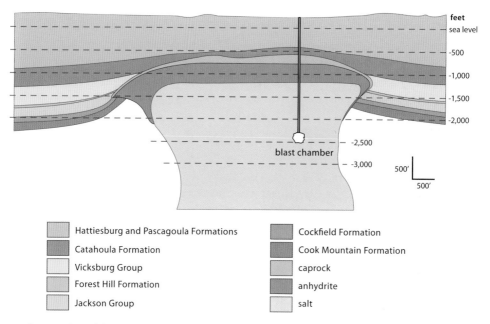

Cross section of the Tatum Salt Dome displaying the borehole, blast chamber, and stratigraphy surrounding the dome. (Modified after Taylor 1972; US Department of Energy 2012.)

INTERSTATE 55
LOUISIANA—MCCOMB—
BROOKHAVEN—JACKSON
93 miles

There are very few rock exposures along this route. From the Louisiana state line nearly to mile marker 50, I-55 travels mostly on gravel, sand, and clay of the Pliocene-age Citronelle Formation. North of here, a few small exposures of Citronelle gravel are visible, such as at exit 61 (visible from the interstate overpass in the southbound lane), near mile marker 65, and just north of mile marker 70. The Citronelle was deposited across southern Mississippi across a broad alluvial plain in a range of environments, from those near the shore to broad braided river channels. Within this stretch of Citronelle, I-55 also travels on terrace and floodplain alluvium of several creeks and rivers: the Tangipahoa River between mile markers 8 and 9, Big Creek near mile marker 31, and a tributary of Bayou Pierre near mile marker 49. Between mile markers 50 and 58, I-55 travels through the Miocene-age clay and sand of the Hattiesburg and Pascagoula Formations, and between mile markers 58 and 61, sand and sandstone of the Catahoula Formation.

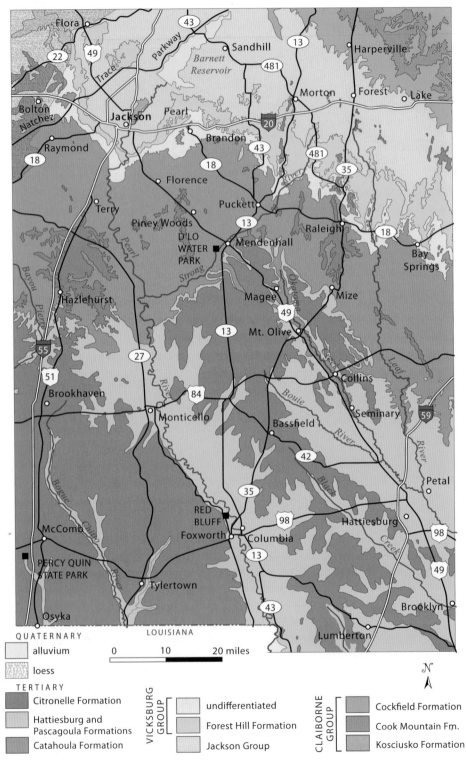

Geology along I-55 between the Louisiana state line and Jackson.

Between mile markers 73 and 84, I-55 travels on Miocene-age sand and sandstone of the Catahoula Formation, deposited on a coastal plain and in deltas as sea level dropped in the Mississippi Embayment. Then between mile markers 84 and 87 it travels on Oligocene-age sand, limestone, and clay of the Vicksburg Group, deposited during a transgressive event. Sedimentation changed from coastal sands (Forest Hill) to carbonate shelf and reef (Mint Spring Formation, Marianna Limestone, Glendon Limestone). The Byram and Bucatunna Formations were deposited as water levels slowly fell. Between mile markers 87 and 88, I-55 travels on sand of the Forest Hill Formation. From mile marker 88 to its junction with I-20, I-55 travels on Yazoo Clay of the Jackson Group in the Jackson Prairie physiographic province. The Yazoo was deposited at the end of a transgressive episode, before sea level began dropping in the embayment. Pleistocene-age pre-loess alluvial terraces occur on top of some of the highest hills along this portion of the route.

US 49
Hattiesburg—Jackson
90 miles
See the map on page 184.

From its junction with I-59 in Hattiesburg to Seminary, US 49 travels on clay and sand of the Hattiesburg and Pascagoula Formations. From Seminary to the MS 35 junction at Mt. Olive, the road travels mostly along the contact between these Miocene-age formations and sand and sandstone of the Oligocene-Miocene-age Catahoula Formation. Okatoma Creek, which parallels the road to the east, has eroded through the Hattiesburg and Pascagoula down into the Catahoula. The former composes the higher-elevation terrain west of US 49, whereas the latter forms the lower-elevation terrain to the east. Most of the Catahoula in the subsurface is described as sand, silt, and clay, but near-surface exposures have become cemented with silica and are sandstone. Locally the Catahoula has been used as building stone and headstones. The resistant Catahoula sandstone forms ledges and chutes in Okatoma Creek, which is a popular destination for canoeing and kayaking. Some of the drops in the creek are rated as Class I rapids.

From near the MS 35 junction to about 2 miles north of the MS 469 junction at Florence, US 49 travels mostly on Catahoula Formation sand and sandstone, except where it crosses through a 7-mile stretch of Pliocene-age Citronelle Formation sand and gravel, north of Magee, and Quaternary-age alluvium of modern streams. Notable examples of alluvium are the deposits of the Strong River and Dabbs Creek, just north of Mendenhall, and Steens Creek, just south of Florence. The alluvial deposits start and end about 1 mile on either side of the crossings.

An outcrop on the east side of US 49 about 0.5 mile north of the MS 13 junction at Mendenhall is a favorite among geology professors. The 20-foot-tall

Class I rapids along Okatoma Creek. The ledge-forming Catahoula Formation sandstone creates several drops along this stretch of the creek. The Catahoula was deposited in alluvial plains, beaches, and shallow nearshore evironments. —Courtesy of John Thomas Cripps

An outcrop of Citronelle Formation sand just behind the Huddle House restaurant on the east side of US 49 in Magee, about 0.5 mile north of the MS 28 junction. Braided streams deposited the Citronelle on a broad alluvial plain.

exposure of Catahoula Formation sandstone features resistant ledges (about 10 feet below the top), crossbedding, and iron staining known as *liesegang bands,* which form as groundwater evaporates from the exposed surface and precipitates iron. Once through the brush the outcrop is accessible.

Starting about 1 mile north of the MS 469 junction at Florence, US 49 crosses a 2-mile stretch of a Quaternary-age alluvial terrace. Unidentified streams that

D'Lo and the Strong River

The D'Lo Water Park, on the banks of the Strong River north of Mendenhall, hosts an audible phenomenon attributed to geology. The falls at D'Lo sound like someone humming or strumming a harp. Some attribute the sound to air bubbles trapped in submerged fissures and scour pockets of the streambed. The sound intensifies as the water level drops, presumably because there is less water to muffle the sound. Some Native Americans considered this sacred ground. On the banks of the river adjacent to Jaynes Falls they practiced a ritualistic ceremony, in which boys were initiated into manhood.

A ledge in Catahoula Formation sandstone exposed in the Strong River at D'Lo Water Park. The small waterfall is approximately 1 foot tall. High flows erode the weathered joints and deepen them. Potholes are created when a rock becomes trapped in a depression or fracture and is moved around by turbulent water, which has a grinding effect. When the holes are deep they are known as potholes, and when they are shallow they are referred to as scour pockets.

traversed the area deposited this sandy deposit. This sand and gravel is older than the pre-loess deposits discussed in the Delta region. US 49 travels briefly on Forest Hill Formation sand and Vicksburg Group marl and limestone of Oligocene age, the Jackson Group of Eocene age, and Quaternary-age floodplain and terrace alluvium of the Pearl River. The boundary between the alluvium and the Jackson Group is difficult to discern.

US 84
Brookhaven—Leaf River
68 miles
See the map on page 184.

Between Brookhaven and the US 49 junction at Collins, US 84 travels in and out of gravel, sand, and clay of the Pliocene-age Citronelle Formation at higher elevations and clay and sand of the Miocene-age Hattiesburg and Pascagoula Formations at lower elevations. The road also crosses the Quaternary-age terrace and floodplain alluvium of numerous streams and rivers.

At Collins, Okatoma Creek has eroded into the underlying Catahoula Formation of Miocene age. The sandstone of the Catahoula provides for great recreational canoeing and kayaking. Its well-cemented sandstone layers resist

Looking east over the Leaf River valley from the crest of Citronelle Formation sand on US 84. The valley is composed of the older Catahoula Formation. The elevation change is approximately 180 feet.

weathering and erosion, so they tend to form ledges that create waterfalls, rapids, and chutes. Some of the small falls and chutes on Okatoma Creek even rate as Class I rapids. East of Okatoma Creek US 84 crosses into the Leaf River basin; the Leaf River has also eroded down through the Citronelle into the Catahoula Formation.

US 98
McComb—Hattiesburg
74 miles
See the map on page 184.

Between McComb and the MS 35 junction near Columbia, the highway travels mostly on Pliocene-age gravel, sand, and clay of the Citronelle Formation, except at low elevations near streams and rivers, where it crosses Quaternary-age alluvial deposits and, in a couple of areas, the clayey deposits of the Hattiesburg and Pascagoula Formations of Miocene age. Notable stream crossings are the Bogue Chitto, at the Pike-Walthall County line, and Magees Creek, about 0.5 mile east of the MS 27 junction in Tylertown. US 98 also crosses 3 miles of the Pearl River's terrace and floodplain alluvium at Columbia.

The stretch of US 98 between Columbia and I-59 at Hattiesburg is much like the stretch between McComb and Columbia: the road travels in and out of gravel, sand, and clay of the Citronelle Formation and clay and sand of the Hattiesburg and Pascagoula Formations, except where it crosses the Quaternary-age alluvial deposits of streams and rivers. The Hattiesburg and Pascagoula are exposed at relatively low elevations at these stream and river crossings, and the Citronelle at higher elevations, away from the floodplains.

Red Bluff

The Red Bluff escarpment, along MS 587 approximately 9 miles north of Foxworth, is one of the most impressive geological features in the state. Locally referred to as the "Grand Canyon of Mississippi," the bluff is being eroded into the Citronelle Formation on the west valley wall of the Pearl River. MS 587 has been relocated twice due to the advancing erosion—350 feet to the southwest most recently. This natural feature has formed primarily due to the erosive force of rainfall runoff, but springs flowing from the Citronelle itself aid in the bluff's destruction.

Any water in the Citronelle at Red Bluff is sourced from precipitation. The water has two effects: it weakens the cement holding the sediments together through chemical dissolution, and it also adds weight to the exposed face of the bluff, which can aid failure. In all probability the bluff will continue to fail and retreat to the southwest. Any engineering done to stop the failure would probably exceed the cost of relocating the road once again. Sometimes it is just not worth fighting natural processes.

The iron oxide–stained sand and gravel of the Citronelle Formation at Red Bluff.

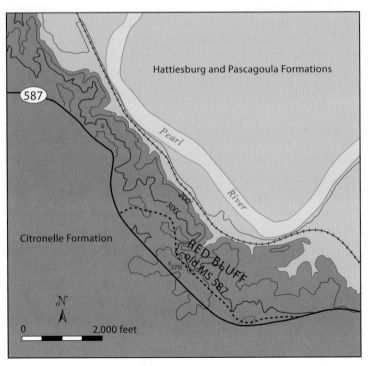

The geology of Red Bluff. You can access the bluff from the old MS 587.

The Citronelle is poorly cemented sand and gravel that was deposited by streams flowing southward over a broad alluvial plain, toward the Gulf, in the late Pliocene and Pleistocene. The sand is composed of quartz, heavy minerals, and feldspar, whereas the gravel is chert, flint, jasper, quartz, and fractured silica known as *tripoli*. Chemical weathering of iron in the sediment has stained it with brilliant purple, red, orange, and yellow iron oxide. The trip to Red Bluff is well worth the time.

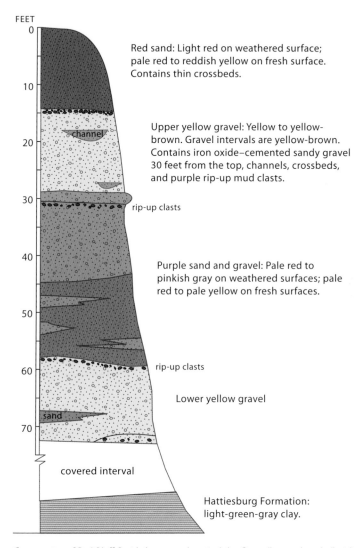

FEET

Red sand: Light red on weathered surface; pale red to reddish yellow on fresh surface. Contains thin crossbeds.

channel

Upper yellow gravel: Yellow to yellow-brown. Gravel intervals are yellow-brown. Contains iron oxide–cemented sandy gravel 30 feet from the top, channels, crossbeds, and purple rip-up mud clasts.

rip-up clasts

Purple sand and gravel: Pale red to pinkish gray on weathered surfaces; pale red to pale yellow on fresh surfaces.

rip-up clasts

Lower yellow gravel

sand

covered interval

Hattiesburg Formation: light-green-gray clay.

Cross section of Red Bluff. Braided streams deposited the Citronelle on a broad alluvial plain. One way geologists can differentiate braided-stream deposits from other channel gravel is the lack of clay deposits. When a meandering stream floods, it leaves behind clay on its floodplain. Rip-up clasts are chunks of material eroded from exposures along the bank and then redeposited downstream. (Modified from Smith and Meylan 1983.)

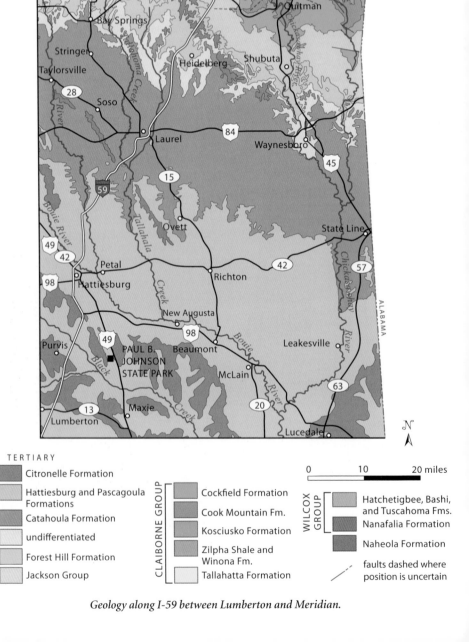

Geology along I-59 between Lumberton and Meridian.

LUMBERTON—HATTIESBURG—
LAUREL—MERIDIAN
112 miles

Between mile marker 41 in Lumberton and mile marker 77, I-59 travels mainly through clay and sand of the Miocene-age Hattiesburg Formation, except where it crosses the Quaternary-age floodplain and terrace alluvium of modern streams. Notable stream crossings are Little Black Creek, near mile marker 48; Black Creek, between mile markers 54 and 55; and the Bouie River, at mile marker 68. Also, between Lumberton and mile marker 63, I-59 goes in and out of gravel, sand, and clay of the Pliocene-age Citronelle Formation at higher elevations. Citronelle exposures are just north of Lumberton (exit 41), just south of Purvis (exit 51), and at the US 98E junction south of Hattiesburg.

A south-facing exposure of the Citronelle Formation on the west side of the southbound lane of I-59 at Hattiesburg, just south of mile marker 63. Braided streams flowing across a broad alluvial plain adjacent to the coastline deposited the Citronelle.

Between mile markers 77 and 112, just south of the Heidelberg exit, I-59 travels through sand and sandstone of the Oligocene-Miocene-age Catahoula Formation, except where it crosses the Quaternary-age alluvium of the Leaf River and Tallahatta Creek. Between mile markers 112 and 121, it travels through the Oligocene-age Forest Hill Formation sand, except between mile markers 118 and 120, where it crosses younger Vicksburg Group sand, limestone, and clay. Around mile marker 121 the highway leaves the Pine Hills physiographic province, and between mile marker 121 and exit 126 at Pachuta it travels on the Eocene-age Yazoo Clay of the Jackson Group. This gently rolling terrain of

Dunns Falls Water Park

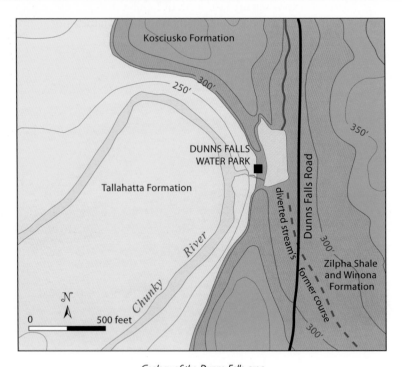

Geology of the Dunns Falls area.

Dunns Falls Water Park features a nineteenth-century geoengineering marvel. In the 1850s John Dunn diverted an unnamed tributary to the Chunky River in order to power a mill. The diversion in the small watershed shortened the tributary's path by 3 miles, but the new discharge point over bluffs, composed of the Basic City Shale Member of the Tallahatta Formation, created a wonderful 65-foot-tall cascading waterfall. Although much of the Basic City is shale, there are abundant siltstone, claystone, and sandstone ledges within it. Abundant silica in the formation was reprecipitated as silica

Dunns Falls flows over the Tallahatta Formation at a historic mill on the Chunky River.
—Courtesy of David Williamson

cement, which makes those units resist erosion better than the underlying shale. The Basic City was deposited in a range of environments, from a nearshore marine shelf to deltas closer to shore. To reach the water park, take exit 142 onto Meehan Savoy Road, and then turn south onto Dunns Falls Road and follow it for about 3 miles.

the Jackson Prairie physiographic province is related to the relative ease with which the clay weathers and erodes.

From exit 126 to just south of the I-20 junction, I-59 travels through the Claiborne Group, part of the North Central Hills physiographic province. For the most part the Claiborne Group sediments are sandy, and thus hilly due to relief caused by their resistance to erosion. I-59 also crosses Quaternary-age floodplain and terrace alluvium, such as just north of mile marker 130, where it crosses deposits of Souinlovey Creek. Near mile marker 131, I-59 passes through a pre-loess alluvial terrace, a remnant Quaternary-age deposit of an unknown stream. All told, there are very few exposures along this portion of I-59.

US 45
ALABAMA—MERIDIAN
75 miles
See the map on page 192.

From the Alabama state line to just north of the MS 145 junction north of Waynesboro, US 45 travels through limestone and marl of the Chickasawhay Limestone and Paynes Hammock Formation and sand, sandstone, silt, and clay of the Oligocene-Miocene-age Catahoula Formation. The Chickasawhay Limestone and Paynes Hammock Formation were deposited farther offshore, but at the same time as the earliest Catahoula Formation. The units are thin and not differentiated on maps, and both are truncated to the west by Catahoula sand. The road then transitions into Oligocene-age Vicksburg Group sand, limestone, and clay to near the MS 145 junction just south of Shubuta. This portion of the route is in the Pine Hills physiographic province.

For about 6 miles from the MS 145 junction, US 45 travels through gently rolling terrain of the Jackson Prairie physiographic province, underlain by the Eocene-age Yazoo Clay of the Jackson Group. The clay erodes easily, leading to the lack of relief in the landscape.

US 45 then transitions into the North Central Hills physiographic province, which it passes through all the way to its junction with I-20 at Meridian. Most of the route passes over the Eocene-age Claiborne Group. The Zilpha Shale and Cook Mountain Formation are clay rich but very thin relative to the Tallahatta, Winona, Kosciusko, and Cockfield Formations. The abundant sandstone, siltstone, and sand in these formations are more resistant to erosion, which leads to more relief. The junction of US 45 with I-20 is on sandy sediment of the Wilcox Group of Paleocene-Eocene age. This sand resists erosion as well, leading to hillier terrain.

US 45 also crosses the Quaternary-age floodplain and terrace alluvium of several modern streams. Notable stream crossings are at Bucatunna Creek, about 7 miles north of the Alabama state line, and the Chickasawhay River, at Shubuta and again near Quitman.

About 1 mile north of the MS 145 exit for Meridian, an eastern face of the Buhrstone Cuesta comes into view. The combined effects of southwest-dipping strata, the ridge-forming characteristics of the Tallahatta Formation's Basic City Shale Member, and the erodibility of the underlying Meridian Sand has resulted in the Buhrstone Cuesta. A cuesta is an asymmetrical ridge that forms due to differential erosion. Because of the regional dip, the cuestas here have gently dipping western slopes and steep eastern slopes. The generally well-cemented Basic City Shale is more resistant to erosion than the underlying Meridian Sand, which is why the shale stands out in relief.

The Buhrstone Cuesta is actually composed of several small cuestas that have been grouped together. A *buhrstone* is a tough, silica-rich rock that was used to make millstones. The Tallahatta Formation was informally known as the "Buhrstone" or "Siliceous Claiborne." It was formally defined and named in 1897, but the cuesta still bears the original name.

An oyster reef exposure in the Cook Mountain, 0.5 mile south of the MS 511 junction in Quitman. The large oysters are Cubitostrea sellaeformis.

Clark County Faults

Faults are generally not visible in outcrops in Mississippi. In Clark County, however, several faults of the Pickens-Gilbertown fault system are exposed. The Pickens-Gilbertown is one of several fault systems that delineate the northern limit of the Mississippi Interior Salt Basin. In Middle and Late Jurassic time, the Gulf of Mexico opened as the continent rifted apart. Tensional forces related to this pulling apart resulted in a system of normal faults that arc to the northwest through the state. A significant amount of salt was deposited in the Gulf as seawater invaded the rifted region, from Florida to Texas, including southern Mississippi's salt basin.

Since the 1970s there have been a number of earthquakes in Clark County, ranging between magnitude 3.0 and 3.5. These earthquakes may be the result of movement along the faults.

At Quitman, immediately south of the MS 511 junction, US 45 drops down off the Cook Mountain Formation and crosses the eroded scarp, or surface expression, of

Looking south at the eroded scarp of the Quitman Fault from the MS 511 junction on US 45. The hill composed of the Cook Mountain Formation is the weathered scarp; from the highway crossover to the photographer's position is the block of Cockfield Formation that dropped down along the fault.

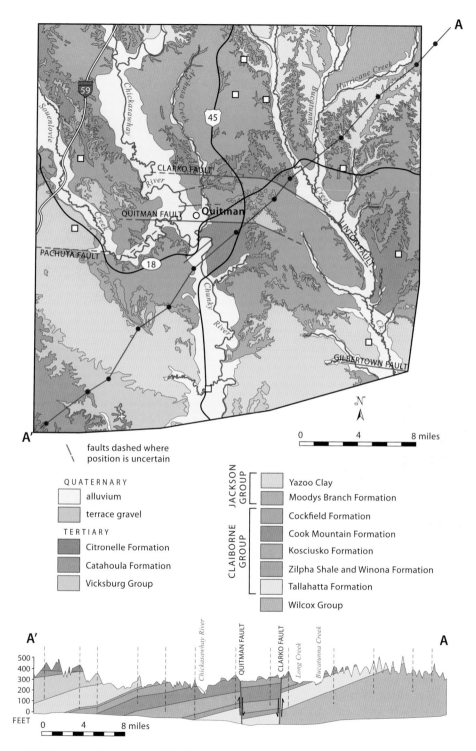

Geologic map and cross section of Clark County. The white squares indicate earthquakes detected between 1977 and 2000. The zone between the Quitman and Clarko Faults has dropped down relative to the areas to the north and south; the displacement of the strata on the the two faults is 55 feet and 120 feet, respectively. The strata dip an average of 30 feet per mile to the southwest, into the Mississippi Interior Salt Basin. (Modified after Gilliland and Harrelson 1978; Bograd 2014.)

the Quitman Fault, passing onto Cockfield Formation sand. The Cockfield Formation dropped down along the fault relative to the Cook Mountain Formation.

The contact between the Kosciusko Formation (red-brown) and the overlying Cook Mountain Formation (tan) on US 45, about 2 miles north of the MS 18 junction. The Clarko Fault lies 0.25 mile to the south, on US 45, where the surficial sediments are primarily those of the younger Cockfield Formation. Cook Mountain sediments are only exposed where streams have eroded down into them. The units south of the fault have dropped down relative to what is pictured here.

US 49
MAXIE—HATTIESBURG
24 miles
See the map on page 192.

Between Maxie, near MS 13, and Hattiesburg, US 49 travels mainly through clay and sand of the Miocene-age Hattiesburg and Pascagoula Formations, except where it crosses the Quaternary-age alluvium of modern streams. However, about 16 miles north of the MS 13 junction, the highway goes in and out of gravel, sand, and clay of the Pliocene-age Citronelle Formation at higher elevations.

Approximately 1 and 4 miles north of the US 98E junction, US 49 crosses through the terrace gravel of Quaternary-age that overlies the clayey Hattiesburg and Pascagoula Formations. Streams eroding into the Pascagoula and Hattiesburg Formations deposited this sandy alluvium.

An approximately 9-foot-tall exposure of a terrace (poorly consolidated yellow-brown sand with gravel) overlying the Pascagoula Formation (white clay). This exposure is on the west side of US 49, about 1 mile north of the US 98E junction.

US 84
LEAF RIVER—LAUREL—ALABAMA
61 miles
See the map on page 192.

From the Leaf River to the west bank of the Chickasawhay River near Waynesboro, US 84 travels on sand and sandstone of the Catahoula Formation, except where it passes through a small area of Pliocene-age Citronelle Formation sand and gravel, 9 miles beyond the Leaf River crossing, and Quaternary-age floodplain and terrace alluvium of modern streams. Notable alluvium crossings include that of the Leaf River, Big Creek, Tallahoma Creek, and Tallahala Creek. Rivers and streams deposited the Catahoula on an alluvial plain or in a delta on Mississippi's southern coast during Oligocene and Miocene time. The Catahoula is recognized as a major ledge-forming unit in southern Mississippi, and it is also known for its abundant petrified palm fossils and very hard quartz- and opal-cemented layers.

From the west bank of the Chickasawhay River to the Alabama state line, US 84 travels mostly through sand, marl, and limestone of the Oligocene-age Chickasawhay Limestone and Paynes Hammock Formation and marl and limestone of the Vicksburg Group. The Chickasawhay Limestone and Paynes Hammock Formation were deposited farther offshore but are the same age as the earliest Catahoula sand. Eventually, the Catahoula sand was deposited over these two units. They are thin and not mapped separately in this book.

A 15-foot-tall exposure of the Catahoula Formation in an abandoned quarry on the south side of US 84, about 1 mile east of the I-59 junction.

Starting about 5 miles east of the US 45 junction, however, the road passes through a 4-mile stretch of Catahoula capped by terrace gravel of Quaternary age on the highest hilltops. Stream crossings with notable Quaternary-age floodplain and terrace alluvium are the Chickasawhay River and Bucatunna Creek.

US 98
HATTIESBURG—MS 63
50 miles
See the map on page 192.

From its junction with I-59, south of Hattiesburg, to about 3 miles south of the Greene-George county line, US 98 travels on the clayey Hattiesburg and Pascagoula Formations, except where it crosses Quaternary-age alluvium. There are scattered exposures of the Pliocene-age Citronelle Formation, though overall there are few exposures of anything along this route. The final stretch into Lucedale crosses into the Citronelle Formation. Ancient rivers scouring north

The contact between the Citronelle Formation (reddish sediment) and the underlying Pascagoula Formation (tan clay) is visible in this 7-foot-tall exposure on the north side of US 98, about 1 mile east of the MS 29 junction.

An outcrop of the Citronelle Formation on the north side of US 98, 4.5 miles east of the MS 57N junction.

and central Mississippi generated its sediments, which were deposited across southern Mississippi on a broad alluvial plain near the coastline.

At its junction with US 49, and about 1 mile east of it, US 98 crosses through sandy gravel that overlies the clayey Hattiesburg and Pascagoula Formations. This terrace marks the floodplain of a river that flowed through this region during Quaternary time. The correlation and dating of ancient river terraces is extremely expensive, so not many have been dated.

From about 4 miles east of its junction with US 49 to near its junction with MS 198W, at Beaumont (the second MS 198 junction), US 98 parallels the Leaf River and travels on and off of its terrace and floodplain alluvium until it crosses the river, about 0.5 mile east of the MS 57S junction. The Leaf River alluvium has been mined extensively over the years. The tailings look like white dunes along the floodplain between US 49 and New Augusta.

MISSISSIPPI 15
BEAUMONT—LAUREL—NEWTON
94 miles
See the map on page 192.

From its junction with US 98 to about 10 miles north of the Perry-Jones county line, MS 15 travels through the clay and sand of the Hattiesburg and Pasca-goula Formations. The road then passes through sandstone, silt, and clay of the Catahoula Formation, of Oligocene-Miocene age, to roughly 4 miles north of the MS 18 junction at Bay Springs. For the next 8 miles, MS 15 travels through Vicksburg Group sand, limestone, and clay, and then Forest Hill Formation sand, all of Oligocene age. These formations, part of the Southern Pine Hills physiographic province, were deposited during the last major sea level rise and marine transgression in the Mississippi Embayment.

MS 15 also crosses Quaternary-age floodplain and terrace alluvium along this stretch of road. Notable deposits are those of the Leaf River, about 2 miles north of the US 98 junction; Tiger Creek, about 3.5 miles north of the Perry-Jones county line; Bogue Homa, about 10 miles north of Ovett; Tallahata Creek, at Laurel; and Tallahoma Creek, north of Laurel.

From about 12 miles north of its junction with MS 18 at Bay Springs to about 1 mile north of the Jasper-Newton county line, MS 15 travels through the Jackson Prairie physiographic province on the Yazoo Clay of the Eocene-age Jackson Group. Only a remnant hill of the Forest Hill and Citronelle Forma-tions, 5 miles north of Montrose, breaks up the gently rolling terrain. Talla-homa Creek and its tributaries have eroded and removed most of the Forest Hill and Citronelle Formations from this region, leaving a few scattered rem-nants on hilltops.

From about 1 mile north of the Jasper-Newton county line to the junction with I-20, at Newton, the landscape becomes hillier; it's part of the North Cen-tral Hills physiographic province. MS 15 travels on the sandy Cockfield and

An 8-foot-tall exposure of Citronelle Formation on the east side of MS 15, about 4.5 miles north of Montrose. The sand was deposited by streams flowing toward the coast.

Looking north from the MS 15 bridge over Tallahoma Creek, north of Montrose. The creek bed is composed of Yazoo Clay; from here MS 15 passes through Forest Hill Formation sand to a remnant hilltop cap of Citronelle Formation sand, gaining 140 feet of elevation.

Cook Mountain Formations of the Eocene-age Claiborne Group. The hilltops are Cockfield, and the hill slopes the Cook Mountain. The barrier and delta sands of the Cockfield overlie clay and limestone of the Cook Mountain. The Cook Mountain was deposited near the end of a transgressive event, in shallow water. The Cockfield delta expanded toward the southwest, ending the transgressive event as sea level dropped. The road crosses notable Quaternary-age floodplain and terrace alluvium of Tarlow Creek about 0.5 mile north of Bethel Roberts Road.

An exposure of the Cook Mountain Formation in the southwest corner of the MS 15 and US 80 junction at Newton. Here, the Gordon Creek Shale Member rests on top of the Potterchitto Sand Member (a channel bed). Ophiomorpha fossils, burrows of shrimp that inhabited offshore sands, are visible in the creek bottom. Some of the casts (inset, center top) weather out of the bedrock. The water appears to be orange because iron oxide minerals are precipitating out of groundwater as it flows into the creek and is exposed to oxygen.

COASTAL MISSISSIPPI

Coastal Mississippi includes road logs from both the Southern Pine Hills and Coastal Meadows physiographic provinces. Elevations in the Southern Pine Hills can approach 400 feet, but those in the Coastal Meadows are less than 60 feet. At first glance, the geology is deceptively simple. Its complexity becomes apparent with closer inspection.

The types of sediment deposited in Mississippi over most of the past 90 million years were heavily influenced by changes in sea level within the Mississippi Embayment. The embayment slowly filled with sediment over time, but the geological units record as many as eight major transgressions and regressions. Clay, limestone, and carbonate-rich sediment generally indicate periods of transgression, or sea level rise, while gravel and sand are largely indicative of regression, or a drop in sea level. However, gravel and sand are deposited both at the beginning of a transgression and at the end of a regression, so these are just broad generalizations. At the close of the Pliocene Epoch, 2.6 million years ago, the Gulf shoreline was fairly close to its present position.

The onset of the Pleistocene Epoch, otherwise known as the "ice ages," once again brought changes to the position of the coastline. Fluctuations in sea level during this epoch, between 2.6 million years ago and 11,700 years ago, influenced not only coastal geology but that of all of Mississippi. Many factors influence sea level fluctuations and shoreline changes. Land areas subside or become elevated by structural forces related to tectonic activity. For example, the swelling of the mid-oceanic ridges can force water onto previously exposed coastal lands. Climate and the circulation patterns of ocean currents related to heating or cooling can change the volume of the ocean. Warm water expands, whereas cool water contracts. In the Pleistocene, though, sea levels were most profoundly affected by the advance and retreat of continental ice sheets and mountain glaciers.

During glacial stages, when ice sheets and mountain glaciers formed far to the north, sea level decreased. The lowered sea levels effectively increased stream gradients and, therefore, the erosive power of rivers and streams. Rivers eroded downward into their channels and became entrenched, bringing larger sediment volumes to the coast to be deposited on the newly exposed continental shelf. Stream-parallel river terraces in these incised valleys formed.

During interglacial stages, when the ice melted and sea level rose, the ocean drowned the recently formed shoreline deposits. During periods of high sea level, sediment accumulated in coalescing alluvial plains that paralleled the

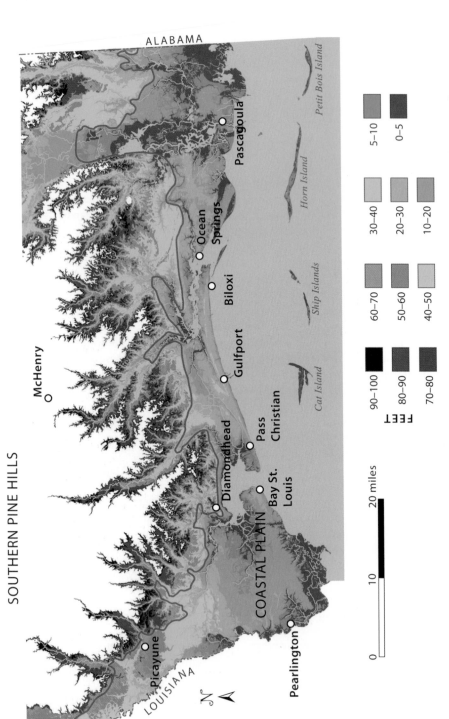

SOUTHERN PINE HILLS

ALABAMA

Pascagoula

Ocean
Springs

Biloxi

Gulfport

McHenry

Diamondhead

Pass
Christian

Bay St.
Louis

COASTAL PLAIN

Picayune

LOUISIANA

Pearlington

Petit Bois Island

Horn Island

Ship Islands

Cat Island

N

FEET		
90–100	60–70	30–40
80–90	50–60	20–30
70–80	40–50	10–20
	5–10	0–5

0	10	20 miles

Digital elevation model of southern Mississippi. The Coastal Meadows physiographic province is south of the red line. In order to highlight how quickly the elevation drops toward the coastline, only elevations up to 100 feet above sea level are shown. The dark-green areas are freshwater and marine-influenced wetlands that are up to 5 feet above sea level.

coastline. When sea level dropped again, these plains were left stranded above sea level as coastal terraces. Because of the slow tectonic uplift of the interior, coastal erosion eliminated all but the two youngest terraces: the Montgomery and the Prairie. Geologists have attributed the uplift and tilting in southern Mississippi, which has elevated the Citronelle, to the addition of sediment (or "loading") in the Gulf of Mexico during Pleistocene time. As the Gulf was loaded, areas along the coast bulged upward in response. Measurements made at the end of the twentieth century revealed uplift rates ranging up to 2 millimeters a year in south-central Mississippi and subsidence rates of up to 1 millimeter per year along the coast.

Scientists understand well the processes responsible for climate change during the Pleistocene Epoch, as well as the Holocene Epoch, which we are in today. Based on climate evidence from ice sheets, sediments, fossils, and other sources, geologists now realize that the Pleistocene wasn't one long ice age. In fact, it comprises many glaciations alternating with warm interglacial intervals of time. Today we find ourselves in an interglacial stage, with global warming, shrinking and disappearing ice sheets and valley glaciers, and shifting climate zones. The most recent glacial stage, the Wisconsin, is the glaciation most relevant to our exploration of Mississippi's coastal geology.

But let's back up a bit. The exposed geology of coastal Mississippi begins in Miocene time with the deposition of the Hattiesburg and Pascagoula Formations, followed by the Graham Ferry Formation in Pliocene time. These units are clay rich and were deposited in environments similar to what you see in coastal Mississippi today (for example, deltas, beaches, dunes, and estuaries). These units are only occasionally exposed in river bluffs and road cuts.

The Pliocene-age Citronelle Formation, laid down in coalescing river floodplains—including braided-stream channels—inland from the contemporary shoreline, veneers much of south-central and coastal Mississippi. The formation consists of sand, muddy sand, silt, and the occasional sandy gravel and clay layers. River-reworked petrified wood, eroded from stream bluffs composed of the Citronelle, are frequently found in southern Mississippi's valleys. Rich assemblages of ancient plant remains and the bright reddish-orange coloration, indicative of mineral oxidation, suggest the Citronelle was deposited in warm-temperate climate conditions, before the climate began cooling in southern North America and sea level began dropping at the end of Pliocene time.

There are four Pleistocene-age units exposed at the surface in coastal Mississippi: the Montgomery Terrace and the Biloxi, Prairie, and Gulfport Formations. Though each of these formations is Pleistocene in age, the Montgomery Terrace is much older than the other three. It is an isolated remnant of an originally widespread coastal plain deposited between 221,000 and 176,000 years ago, during an interglacial stage and a time of higher sea level. The terrace is composed of river-deposited sand and silty sand with a few interlayered lignite beds that contain plant remains and carbonized wood fragments.

The Biloxi, Prairie, and Gulfport Formations were deposited during the Sangamon interglacial, between 132,000 and 107,000 years ago. During this time sea level peaked between 22 and 28 feet above today's level in the Gulf.

Because of the gentle slope of the coastal plain, the shoreline advanced several miles inland. As the Gulf transgressed over the plain, stream valleys became inshore bays and lagoons, filling with fossil-rich sandy clay and sandy mud of the Biloxi Formation; these deposits often contain fossil oyster reefs. Farther inland, the wedge of the Biloxi deposits tapers and interfingers with silty sand, sand, and gravel layers of the Prairie Formation, deposited by braided and meandering rivers on relatively flat floodplains during the Sangamon interglacial and early Wisconsin glacial stages.

The Gulfport Formation developed on top of the Biloxi deposits. With plentiful sand delivered to the shoreline by westward-directed longshore drift, a succession of shore-parallel wave-built beach ridges and wind-built sand-dune ridges formed. These ridges mark each successive advance of the shoreline southward. The end result was a wide beach-ridge plain known as a *strand plain*. The strand plain was built and maintained while sea level remained elevated for a while. Today this sandy barrier-ridge complex stretches a few miles inland from the present shoreline.

Following the peak of the Sangamon interglacial (107,000 years ago), climate fluctuated during a transitional period. The transition between the end of the Sangamon and the start of the Wisconsin was from 107,000 to about 71,000 years ago. Eventually the globe once again cooled, ice built up on land, and sea level declined, ushering in the Wisconsin glacial stage. During the Wisconsin glaciation, the Mississippi shoreline gradually retreated seaward. At the peak of the glaciation, between 21,000 and 19,000 years ago, sea level stood 410 to 450 feet lower than today. The former sea bottom was exposed, and the shoreline was 90 to 100 miles offshore from where it is today. The Gulf of Mexico is fringed by a broad continental shelf. As its name implies, the shelf is a relatively flat surface that sits along the coastline in relatively shallow water. Today, 90 to 100 miles offshore of Gulfport, the shelf sits beneath 500 to 600 feet of water.

During this period of shoreline retreat, Mississippi's rivers continued to deposit sediment to the ever-expanding coastal plain. Declining sea level increased their gradients, causing the rivers to deeply incise their channels. Not only were large rivers, such as the Mississippi, Mobile, Pearl, and Pascagoula, affected, but smaller streams also deepened and widened their channels and valleys.

By about 11,700 years ago (the beginning of the Holocene Epoch), most of the ice covering North America had melted. The climate had warmed significantly and sea level was rising at a rapid rate. The rise had slowed by 6,000 years ago and then even more by 4,500 years ago. All the river and creek valleys on the Mississippi coast were eventually drowned and partially filled with sediment, and the shoreline started to approximate the present one. A barrier island chain formed along the coast and extended from Louisiana into western Mississippi about 4,500 years ago. Approximately 3,600 years ago, the deltas of the Mississippi and Pearl Rivers surrounded the easternmost islands in this chain. In Mississippi today, these former barrier islands, now identified as Point Clear and Campbell Islands in Hancock County, are surrounded by tidal marshlands, not water.

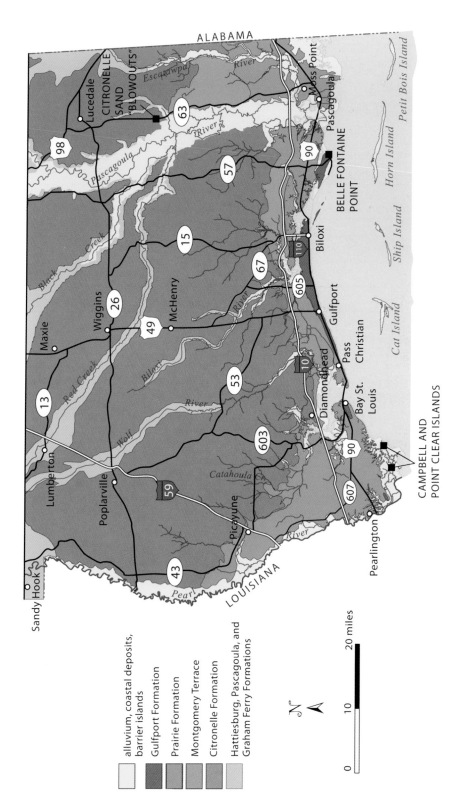

Geology of coastal Mississippi. (Modified after Dr. Ervin Otvos, personal communication.)

ALABAMA

LOUISIANA

Sandy Hook

Lucedale

CITRONELLE
SAND
"BLOWOUTS"

Escatawpa River

Pascagoula River

Moss Point

Pascagoula

BELLE FONTAINE
POINT

Petit Bois Island

Horn Island

Ship Island

Cat Island

Maxie

Wiggins

McHenry

Black Creek

Red Creek

Biloxi River

Lumberton

Poplarville

Picayune

Catahoula Cr.

Wolf River

Pearl River

Pearlington

Bay St.
Louis

Pass
Christian

Diamondhead

Gulfport

Biloxi

CAMPBELL AND
POINT CLEAR ISLANDS

98

63

57

15

26

49

67

605

110

90

13

53

603

59

43

10

90

607

alluvium, coastal deposits,
barrier islands

Gulfport Formation

Prairie Formation

Montgomery Terrace

Citronelle Formation

Hattiesburg, Pascagoula, and
Graham Ferry Formations

N

0   10   20 miles

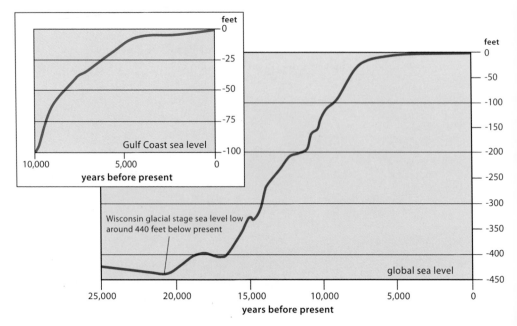

The estimated sea level in the Gulf of Mexico during late Pleistocene and Holocene time. Sea level varies slightly for coastal areas across the globe due to localized uplift or subsidence, the growth or melting of glaciers, and the thermal expansion of seawater. (Modified after Otvos and Giardino 2004; Lambeck et al. 2014.)

BARRIER ISLANDS

Barrier islands are generally elongated, narrow bodies of sand that form along passive continental margins—that is, a continental plate boundary free of tectonic activity, such as earthquakes and volcanic eruptions. Barrier islands are geologically transient and ephemeral features, but they have an important role in coastal systems. Perhaps the greatest function barriers serve is to protect the coast from tropical storms and hurricanes. Though they typically do not rise very high out of the water, their presence can decrease the wave energy from weaker storms that pass over them and reach the coast.

Barrier islands form as a consequence of longshore currents (currents that roughly parallel the coastline), sediment supply, water depth, tidal range, and wave activity. When all of these factors are tuned in just the right way, these long, linear deposits of sand can develop. The islands are geologically ephemeral features that migrate, erode, are destroyed, and re-form whenever there is a change in any of the factors. Catastrophic events such as hurricanes and tropical storms affect barrier islands, moving them, cutting them in two, and sometimes destroying them. Long-term changes in climate and sea level will also affect the islands, causing them to move inland or even making them disappear.

Mississippi's coastline is partially protected by five large barrier islands that are estimated to be between 5,000 and 5,700 years old. From east to west they are Petit Bois, Horn, West Ship, East Ship, and Cat Islands. They are part of the Gulf Islands National Seashore, though some of Cat Island remains in private ownership. Deer Island is not a Holocene-age barrier; it is a remnant of an older barrier-ridge complex of the Gulfport Formation.

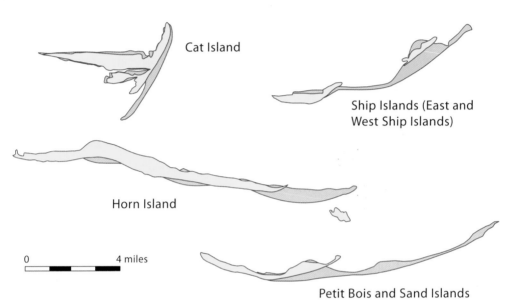

The shape of Mississippi's large barrier islands in 1850 (brown) and 2010 (yellow). Most of the islands have migrated slightly westward, indicating the overall direction of longshore currents in the region. Hurricane activity split Ship Island into East and West Ship Islands. (Modified after Schmid 2000.)

Mississippi's modern barrier islands are on the move and also eroding. Petit Bois and Horn Islands, south of Pascagoula, have migrated westward approximately 2 miles during the last 150 years largely due to wave-driven longshore drift in the Gulf. Cat Island, south of Gulfport, has not migrated and has only eroded in place. Ship Island, just east of Cat Island, has both migrated and eroded, but more significantly it was cut in half by Hurricane Camille. Fort Massachusetts, originally constructed on the western tip of the island between 1859 and 1866, is now situated approximately 1 mile from the island's western tip because of the westward buildup of sand.

Mississippi's barrier islands have suffered an accelerated rate of erosion since the mid-1800s, and especially since the mid-twentieth century. Scientists attribute this erosion to reduced sediment supply and frequent strong storms. Since the 1850s Petit Bois Island has lost 43 percent of its sand, Horn Island 26 percent, Dog Island (or Isle of Caprice) 100 percent (more on this later), Ship Island 57 percent, and Cat Island 40 percent.

In addition to the natural changes in sediment dynamics, the creation of ship channels has significantly impacted the migration of the islands. The Gulfport, Biloxi, and Pascagoula ship channels are located off the western side of Ship, Horn, and Petit Bois Islands respectively. Sand eroded from the eastern part of the islands is carried westward and lost in the ship channels. Without the sediment-trapping ship channels, the islands would migrate unimpeded and be better able to survive. (Interestingly, Sand Island, west of Petit Bois, is a small barrier island that was constructed from sand dredged from a channel to keep it passable.) The large-scale relocation of shipping channels is prohibitively expensive, but some scientists calculate that the loss of the barrier islands could prove even costlier. Without them, Mississippi's coast would be much more vulnerable to storm and hurricane damage.

East and West Ship Islands were once connected. However, in 1969 Hurricane Camille washed over and removed enough sand to divide the island in two. Hurricane Katrina widened the gap between the islands even more, perhaps permanently. In an effort to maintain the barrier system, the US Army Corps of Engineers has a tentative restoration plan underway to reconnect the two ends of Ship Island. The project will pump sand into the 6-mile-long and nearly 6-foot-deep gap—known as Camille Gap—between the two islands. As of 2015 the project had not yet started.

Except for two of them, Mississippi's barrier islands have rarely been populated. Cat Island, visited by prehistoric oyster gatherers, has been inhabited since the late 1870s and boasts a rich history that includes pirates, gangsters, Seminole Indians, and US military training exercises during World War II. Ownership is currently divided between private individuals, the state of Mississippi, British Petroleum, and the US government. In a bold move, entrepreneurs developed a

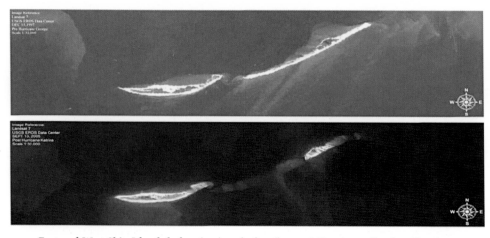

East and West Ship Islands before (top) and after (bottom) Hurricane Katrina. Hurricane Camille first cut the island in half in 1969, but Katrina increased the separation considerably.
—Courtesy of the US Geological Survey

casino and hotel on Dog Island (or Dog Key), west of Horn Island, in the 1920s, and renamed it Isle of Caprice. By 1932 shifting sand had destroyed the development. Today the shoals at Dog Keys Pass, between Ship and Horn Islands, mark the site of the former island and its doomed development. Needless to say, small barrier islands are no place for permanent structures.

COASTAL MODIFICATION BY HUMANS

Mississippi's mainland shoreline is flanked by a "low-energy" beach. This designation means that significant changes only take place during major storm events. Human interest in water and economic development are the two reasons the coast is so heavily modified. Coastal scientists have suggested that, when accounting for the seawalls and artificial beaches in Hancock and Harrison Counties, Mississippi may have one of the most engineered coasts in the United States. Presently, the Mississippi coast is 57 percent engineered and 43 percent natural.

Early in the coast's history of human development, it was known for oyster and seafood canneries. The first cannery was established in 1880, and by 1901 Biloxi was second only to Baltimore in domestic seafood production. Although Mississippi was not known for its beaches, it did have narrow sand beaches, and people were drawn to the water. As early as the late 1800s, people were utilizing the 80-to-100-foot-wide beach as a road. As buggies and early automobiles became more common in the region, the road surface was filled with whole, and later crushed, oyster shells. The road, which followed a path that included Pleistocene dunes, beaches, and back-beach areas, became known as Shell Road or Shell Drive. Despite serious damage to Shell Drive by hurricanes in 1893, 1901, 1909, and 1915, the beach "highway" crossed nearly the entire length of Harrison County and was protected by the Mississippi Legislature. Eventually this highway was integrated into the Old Spanish Trail, a southern route that spanned between Florida and California. Today US 90 follows much of this old route in Mississippi.

Gulfport Harbor was one of the earliest alterations of the Mississippi coastline. Constructed after the hurricane of 1893 to create a safe harbor for ships, it did little to protect the coast from erosion and damage. Coastal developers and residents endured storm damage well into the early 1900s.

The 1915 hurricane caused considerable damage to the coast and prompted local officials to request state intervention to protect increasing coastal investments and growing infrastructure. The result was the planning and development of a 26-mile-long, 8-to-11-foot-tall stepped seawall to make the Harrison County coast more aesthetically pleasing for tourists and to protect the coast from erosion during storm events. This seawall, Mississippi's first large-scale coastal alteration, was completed in 1928.

Shortly after the seawall was built, natural beaches began to disappear despite local efforts to sustain them. By the late 1930s it was common to see the waters of Mississippi Sound lapping at the base of the recently constructed seawall. The famed Broadwater Hotel, in Biloxi, created Mississippi's first artificial beach in 1940 by pumping sand between two groins (artificial barriers constructed perpendicular to the coastline). This first beach nourishment project was done to ensure that hotel patrons had a beach to use, not to save infrastructure along the coast. A 1944 federal study suggested that an artificial beach would reduce the potential failure of the seawall in significant storms by reducing the strength of the waves before they hit the wall. In 1947 a hurricane did destroy much of the Biloxi beachfront and several sections of seawall, but the artificial beach in front of the hotel suffered only minor erosion, and the wall there was left intact.

The US Army Corps of Engineers began adding sand to Mississippi's beaches, and by 1952 the corps had completed the world's longest artificial beach, known as the Harrison County Sand Beach, which extended from Henderson Point east to the Biloxi Lighthouse. This 25-mile-long beach ranged up to nearly 300 feet wide, and it's estimated that it required approximately 4.5 million cubic yards of sand. Following Hurricane Betsy in 1965, a portion of unprotected seawall in Hancock County sustained severe damage. By 1967 the corps had completed a 6-mile-long beach in front of this section in order to protect it.

Though all of Mississippi's artificial beaches were in need of additional sand (nourishment) before Hurricane Camille in 1969, funds were not appropriated until 1973. They were again nourished in 1985, 2001, and 2008. The last three beach nourishment projects averaged 0.8 million cubic yards of sand each.

Undoubtedly, the Mississippi coastline would look very different without the large-scale geoengineering efforts of the last century. Not all areas of the coast are artificially protected. The unprotected coastal wetlands along the shore are subject to extreme wave erosion. The greatest rates of erosion occur

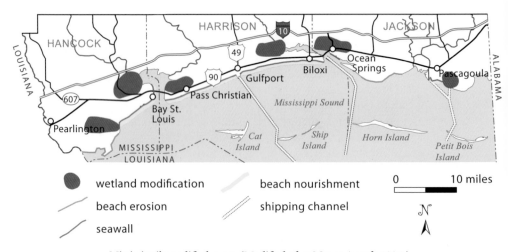

Mississippi's modified coast. (Modified after Meyer-Arendt 1994.)

A portion of seawall built in 1951 (after nourishment had been done) on the west shoreline of Bay St. Louis, as seen from just north of the US 90 bridge. The lowermost platform is a more recent addition to the old wall.

in the natural wetlands of western Hancock County. Between 1850 and 1986 an average 5 feet of wetland was lost per year. There is no natural sand in the area, and the exposed clay-rich marsh sediment is undercut by waves; eventually it slumps off and is carried away. Sand resources appropriate for beach nourishment in this area are not readily available offshore.

Coastal modifications also take place inland. To make low-lying wetland areas suitable for development they typically have to be elevated above the groundwater level with artificial fill. Wetland modification requires a federal permit from the US Army Corps of Engineers that is granted only after the Mississippi Department of Environmental Quality has approved of the modification.

Despite human efforts such as seawall construction, beach nourishment, and the infilling and development of beachfront and wetland areas, the Mississippi coast has lost 3,150 acres to erosion since 1850. During that same period, 1,725 acres were added through coastal development, for a net loss of 1,425 acres.

INTERSTATE 10
LOUISIANA—GULFPORT—BILOXI—
MOSS POINT—ALABAMA
77 miles
See the map on page 211.

Between the Louisiana state line and the Alabama state line, I-10 travels mostly through the low and flat Coastal Meadows physiographic province. Between the state line and mile marker 1, I-10 travels on sand, silty sand, mud, and clay of the Holocene-age tidal marsh of the Pearl River delta. The Pearl River is the largest and longest river in Mississippi. Its wetlands are the habitat of the endangered Gulf sturgeon (*Acipenser oxyrinchus desotoi*) and possibly the ivory-billed woodpecker (*Campephilus principalis*). From mile marker 1 to exit 16, the road travels on the Prairie Formation, except where it traverses marshland and tidal channels of the Jourdan River, near mile marker 14. The Prairie Formation is composed of sand, silty sand, mud, clay, and occasional gravel deposited by braided and meandering rivers during the Sangamon interglacial and early Wisconsin glacial stages.

From exit 16 to just east of mile marker 26, I-10 travels across sand, silty sand, mud, clay, and occasional gravel deposits of the Citronelle Formation. The greatest elevations on the coast are at Diamondhead, immediately

View across intertidal marsh and tidal channels of the Holocene-age Jourdan River delta, south of the I-10 bridge near mile marker 14.

northeast of exit 16. The Citronelle ranges up to 100 feet above sea level here. These Citronelle hills, and those to the north that rise up to 420 feet above sea level, were preserved because of the regional uplift of southern Mississippi during the Quaternary; the increased gradient—the result of uplift—caused rivers and streams to cut downward and become entrenched as they flowed to the coast. As a result, they didn't remove the remnant hills. There are numerous sand and gravel mines in the area that take advantage of these commercially valuable deposits.

From just east of mile marker 26 to exit 44, I-10 travels mostly on the level Prairie alluvial surface. Exceptions are where it crosses floodplain and terrace alluvium of modern streams (the Wolf River delta and tidal marshes just east of mile marker 27, Bernard Bayou just east of exit 34, the Biloxi River near mile marker 39, and the Tchoutacabouffa River near mile marker 43) and the Citronelle Formation (between exit 31 and mile marker 32, and just west of mile marker 41). From exit 44 to just east of exit 56, I-10 travels on the Pleistocene-age Montgomery Terrace, a sandy coastal plain that formed 221,000 to 176,000 years ago, when sea level was higher.

Big Ridge Escarpment

Between exit 44 (Cedar Lake Road) and exit 57 (MS 57), 0.5 mile south of the interstate, lies an interesting feature known as the Big Ridge Escarpment. This 13-mile-long, nearly linear feature stretches between the Tchoutacabouffa River on the west and Old Fort Bayou on the east. The change in elevation from the top of the escarpment to its toe is 15 to 20 feet. Two theories have attempted to explain the origin of the escarpment. The first theory maintains that wave erosion formed the scarp. However, there are no marine deposits at the scarp, and given what we know of the region's geologic history, there isn't a past sea level high-stand that could have cut the scarp.

The second theory maintains that normal faulting formed the escarpment: at some point in the past, the ground in front of the scarp (the fault block) dropped downward relative to the ground behind the scarp (the up-thrown fault block), and the scarp itself is the surface expression of the fault. This origin is supported by the escarpment's linear, slightly arcuate trend and the abrupt but continuous elevation difference across its entire length. Aerial photos taken before urban development corroborate the fault theory. The photos show rectangular drainage patterns etched into the land surface in the up-thrown block, parallel and perpendicular to the escarpment, and linear ponds (known as sag ponds) parallel to the base of the scarp. These features would have formed as a direct result of faulting as the down-dropped block rotated into the fault as it sank. The ponds no longer exist.

The Big Ridge Escarpment marks the contact between the Montgomery Terrace and the Prairie Formation and may be a fault scarp.

To see the escarpment, take exit 44 south 0.35 mile to Popps Ferry Road. The escarpment closely parallels the east-west Popps Ferry Road along its southern edge. There's a lot of development, but you can note the 10-to-15-foot drop. North of the escarpment the ground is composed of the Montgomery Terrace, white and yellow-brown silty sand and sand that often contains peat layers with carbonized wood fragments. The sediments are part of a coastal plain that formed during an interglacial period with high sea level. South of the escarpment is the Prairie Formation.

From just east of exit 56 to near exit 57, I-10 crosses Holocene-age flood-plain and terrace alluvium of Old Fort Bayou. From there to the Alabama line, I-10 travels on and off the Prairie Formation. Between mile markers 63 and 68 it passes through a wide stretch of the Pascagoula River delta and related tidal marshlands. Between mile markers 72 and 74 the interstate crosses Black Creek and then dense swamp forests fringing the Escatawpa River.

Roughly 2.5 miles south of I-10, the sizeable Escatawpa makes a sudden 90-degree westward turn, north of US 90 at Moss Point, before discharging into the Pascagoula River. The abandoned channels and oxbow lakes east of Moss Point, located in Holocene-age wetlands, mark the position of the Escatawpa River prior to its capture by the Pascagoula River. The fast-shrinking, abandoned remnants of the Escatawpa delta, where the river once discharged into the Gulf, are protected in the Grand Bay National Estuarine Research Reserve, located on the shores of Grand Bay, southeast of Moss Point.

Pascagoula River

The Pascagoula River is also known as the "Singing River." The name is based on a legend describing the peace-loving Pascagoula Indians, who walked hand in hand into the river while singing to avoid fighting the invading Biloxi Tribe. The Pascagoula River is 80 miles long and has a drainage basin that encompasses nearly 9,000 square miles. Based on rivers with similar discharge, it is the largest river in the lower forty-eight states to remain in its natural state.

A great deal of conservation effort has been focused on the Pascagoula. Conservationists fought a US Department of Energy proposal to create a cavity in the Richton Salt Dome for the Strategic Petroleum Reserve. It's estimated that the project would have required 50 million gallons of the river's water per day over five years to dissolve the salt, creating space for 160 million barrels of oil. The project stalled in 2011 due to environmental concerns for wetlands downstream of the project site, which would have been impacted by the construction of over 330 miles of pipelines and by the reduced volume of water supplied to them.

INTERSTATE 110
US 90—I-10
4 miles
See the map on page 211.

I-110 is entirely within the Coastal Meadows physiographic province. From its junction with US 90 to near mile marker 1, I-110 travels on the Gulfport Formation, a complex of ocean-wave and wind-deposited beach ridges that formed during the Sangamon interglacial, when sea level was 20 feet higher and the coastline several miles inland. From there the road travels on the Prairie Formation for the next 2.5 miles, except where it crosses the Back Bay of Biloxi. The Back Bay is actually a valley the ancient Biloxi River cut when sea level was low and the river's gradient thus steeper. Marsh islands, visible from the bridge over the Back Bay, outline the original, subsequently drowned meandering river channel.

The Prairie Formation is composed of sand, clay, and gravel deposited by rivers on a relatively flat floodplain as sea level fell during the transition from the Sangamon interglacial to the Wisconsin glacial stage. In places the Biloxi Formation is visible beneath the thin Prairie Formation along Biloxi Bay and the Industrial Waterway. The Biloxi is composed of estuarine, nearshore, and offshore sand and mud deposited along and south of the Sangamon coastline. As sea level dropped, the Prairie was deposited over it.

The interstate then crosses the sandy Montgomery Terrace for about 0.5 mile to its junction with I-10. This terrace was a coastal plain that developed between 221,000 and 176,000 years ago, during an interglacial time that predates the Sangamon interglacial.

US 49
GULFPORT—MAXIE
49 miles
See the map on page 211.

In Gulfport, US 49 starts out in the Coastal Meadows physiographic province. For 1 mile, from its junction with US 90 to Brickyard Bayou, it crosses a belt of ancient beach and dune ridges of the Gulfport Formation. This beach ridge plain, or strand plain, grew toward the Gulf during the late Pleistocene. The oldest ridges are those that formed along the plain's northern margin, while the youngest ridges developed along the southern margin, where the present Mississippi Sound beach is located.

Between Brickyard Bayou and I-10, US 49 crosses a flat surface underlain by the Prairie Formation, composed of silty sand with occasional gravelly sand and clay beds. Rivers and creeks deposited the sediment in their channels and on adjacent floodplains during the transition from the Sangamon interglacial stage to the early Wisconsin glaciation. During the Sangamon interglacial, sea

Looking north at a low-relief escarpment on US 49, just north of Bernard Bayou at the I-10 junction in Orange Grove. The change in elevation beneath the stoplights corresponds with the contact between the level Prairie Formation, to the south, and hilly terrain underlain by the Citronelle Formation, to the north.

An 8-foot-tall exposure of Citronelle Formation sand on the west side of US 49 at the Stone-Forrest county line, 1 mile north of Bond. Rivers emptying into the Gulf deposited the Citronelle on floodplains. The brilliant orange-red color is due to the oxidation of iron in the sediment.

level gradually declined and streams and creeks extended their courses toward the Gulf, eventually reaching the shoreline of the lowest Pleistocene sea level 90 to 100 miles offshore. Large areas of today's shallow continental shelf, therefore, became part of an expanded Pleistocene coastal plain.

North of the junction with I-10 the highway crosses into the Southern Pine Hills physiographic province where the road reaches the southern limit of the Pliocene-age Citronelle Formation. The post–ice age regional uplift of the vast southern Mississippi region that is covered with the Citronelle explains the northward-rising, undulating hill country north of I-10. The uplift has caused streams and rivers to cut their channels deeper, creating the steeper slopes of the Citronelle topography. The road also crosses floodplain and terrace alluvium of streams, including that of the Biloxi River, about 4.5 miles north of the MS 53 junction, and Red Creek at Perkinston, about 9 miles north of the Harrison-Stone county line. At lower elevations near the stream and river crossings—for about 1 mile on either side of the Biloxi River and for about 3 miles on either side of Red Creek—US 49 travels on the clayey Hattiesburg and Pascagoula Formations of Miocene age, which are covered by silty sand of the Citronelle Formation at slightly higher elevations.

US 90
Louisiana—Gulfport—Biloxi—
Pascagoula—Alabama
86 miles
See the map on page 211.

In 1925, the Old Spanish Trail along the Mississippi shoreline was designated US 90. Between the Louisiana and Alabama state lines, US 90 travels entirely through the Coastal Meadows physiographic province, composed of sand, clay, and gravel of Holocene and Pleistocene age. The Pleistocene deposits are alluvial, deposited by several rivers as they crossed the coastal plain toward the Gulf, whereas the Holocene deposits are deltaic.

From the Louisiana state line to about 2 miles east of the east shore of St. Louis Bay, US 90 travels predominantly on the Pleistocene-age Prairie Formation. This formation, mostly sand and silt, occasionally sandy gravel, and a few clay layers, was deposited by multiple rivers during the transition from the Sangamon interglacial stage to the early Wisconsin glacial stage.

South Beach Boulevard, south of US 90, runs on the Prairie surface along the Mississippi Sound shores of Bay St. Louis and Waveland. A low seawall faces a narrow, eroding, occasionally nourished beach along the highway to road's end at the Silver Slipper Casino, near Point Clear Island. The southern

Grand Bayou in the Grand Bayou Coastal Preserve, southwest of Waveland. This marshland and bayou developed in an ancient lagoon that existed between two Holocene-age barrier islands (Point Clear and Campbell) and the mainland.

portion of Hancock County is composed of marshland, tidal channels, and relict barrier islands related to sea level fluctuations during the Holocene. Approximately 4,500 years ago, a line of barrier islands and lagoons landward of the islands evolved between Alabama and southeastern Louisiana. Approximately 3,600 years ago, sediment from the deltas of both the Mississippi and Pearl Rivers surrounded the barriers in Mississippi, creating tidal marshlands. Here, Point Clear and Campbell Islands are these relict barriers. Under New Orleans, to the west, delta deposits completely buried the islands of this chain.

From about 2 miles east of the east shore of St. Louis Bay to the west shore of Biloxi Bay, US 90 travels on the southern margin of the Gulfport Formation, a remnant of the mainland strand plain of the Sangamon interglacial stage. The strand plain developed when the shoreline was north of the present one and sea level was 22 to 28 feet higher than today. A semicontinuous dune ridge, located along the edge of the level Prairie Formation alluvial plain, marks the oldest, most northern Sangamon-age shoreline. In this area, this ancient shoreline is along the south shore of the Back Bay of Biloxi. As the strand plain advanced toward the Gulf, numerous parallel ridges and swales (valleys) developed parallel to the coastline. The youngest of the Gulfport beach ridges occurs just inland of the present Mississippi Sound shoreline. This subdued ridge and swale topography, which marks the seaward growth of coastal land, is clearly visible in numerous north-south-oriented streets, perhaps most prominently along Beauvoir Road between US 90 and Pass Road in west Biloxi.

This portion of US 90 parallels one of the longest artificial beaches in the world. Completed in 1952, Harrison County Sand Beach is composed of white quartz sand that was mined as far as 1,500 feet offshore and pumped, via floating pipelines, to the coast. It's estimated the project required 4.5 million cubic yards of sand. Major beach nourishment took place in 1973, 1985, 2000, and 2001. The rate of beach erosion is increasing, which reduces the length of time between necessary nourishments. The primary factor affecting beach erosion appears to be climate change and the resulting increased sea level.

Beach nourishment is not unique to Mississippi. Many coastal states recognize the value of the beach when it comes to protecting coastal infrastructure and attracting tourists. Between 1950 and 2006 the US Army Corps of Engineers nourished more than 360 miles of beach on the Atlantic and Gulf coasts. Beach nourishment is costly, both in terms of dollars and effects on the environment. As federal dollars have become increasingly difficult to obtain, community leaders in Mississippi's three coastal counties have looked to alternative solutions for maintaining their beaches.

Research conducted in the mid-2000s documented that the practice of raking and combing the sand with heavy equipment was partially to blame for erosion. Raking fluffs the sand, removes large debris, and makes the beach more aesthetically pleasing, but the heavy equipment compacts the beach's lower layers. As a result, the sand on top is more easily blown away. It's also more easily washed away because the compacted layers inhibit water infiltration, which promotes storm-water runoff. As a result, sand is washed into the Gulf, where it is taken away by longshore drift.

In some areas, bioengineering techniques that use natural vegetation to reduce erosion and maintain the beach are being evaluated. Coastal infrastructure is dependent on the protection offered by artificial beaches, and the beaches must be maintained, but large-scale nourishment programs are expensive, and it's difficult to find funding for them. The hope is that large-scale bioengineering projects will maintain the beach effectively and will also be financially sustainable.

While driving along the coast you can't help but notice the majestic southern live oaks (*Quercus virginiana*). Live oak trees are adapted to thrive in salty soils. A very deep taproot permits the trees to access freshwater, and the tree's low center of gravity makes it very stable in hurricane-force winds. Both attributes make it a long-lived tree on the coast.

Although the nourished beach was added to protect the coast, it behaves just like any natural beach. Along the Mississippi coast the prevailing waves and currents move from east to west. The beach migrates westward in the same manner as the barrier islands. Waves and wind generally strike the coast along a northwesterly trend. Sand grains are washed onto the beach at a northwesterly angle and then retreat straight back into the surf, creating a zigzag pattern that slowly moves the sand grains westward. This process is called longshore drift. Numerous concrete storm-drain pipes stretch across the Harrison County Sand Beach to the Mississippi Sound and act as obstructions that locally

The allegedly 500-year-old Friendship Oak (a southern live oak) on the Long Beach campus of the University of Southern Mississippi.

Harrison County Sand Beach is one of the longest engineered beaches in the world. It is composed of white quartz sand and was originally created with sand pumped from offshore areas. The grassy ridge in the upper-left corner is the youngest beach ridge of the Gulfport Formation strand plain.

A storm drain extending into the Mississippi Sound. Sand accumulates on the east side of the pipe while erosion removes it on the west side. The larger the obstruction, the wider the beach extension on the wave-dominated side. The difference in the coastline width in this image is approximately 20 feet.

disrupt longshore drift. Stop at any one and you will find that sand piles up against the eastern side of the pipe, extending the beach toward the Gulf, whereas the sand on the pipe's western side has been scoured away.

Sand dunes form along coastlines when plant species adapted to growing in the salty and sandy environment take hold. As vegetation expands, the dunes grow. Along Mississippi beaches sand fencing is used to help the dunes become established. The dunes along the Harrison County Sand Beach vary from newly planted areas to more mature dunes that support pine and palm trees and attain a height of up to 10 feet or so. All the dunes on this beach have been planted with native vegetation, including the beautiful tasseled sea-oat grass. The plantings were designed to replicate the sand-accumulating function of natural sand dunes. Dunes reduce wind erosion, protect the highway from dangerous windblown sand, and may briefly and slightly reduce wave action. They also provide habitat for nesting shorebirds, such as the least tern (*Sternula antillarum*). During the breeding season, sections of the beach are closed to minimize disturbances to the terns.

The narrow, spindle-shaped Deer Island is visible across a narrow inlet south of Biloxi. In the past century this barrier island underwent major erosion. The recent construction of artificial marsh on its southeastern end, however, restored much of the lost land area. The island's northwestern sector is underlain by remnants of the late Pleistocene Gulfport Formation, and the southeastern sector is composed of Holocene-age and modern sand deposits.

Dunes created with introduced vegetation on the Harrison County Sand Beach.

Deer Island is visible across Biloxi Bay from the eastern Biloxi Peninsula. Although still standing, much of the pine forest is dead, killed by a combination of storm winds, storm-related salt poisoning of the soil, and the long drought that followed Hurricane Katrina in 2005.

Biloxi Lighthouse

The Biloxi Lighthouse, built in 1848, is in the US 90 median just west of the I-110 junction. It was one of the first cast-iron towers in the South, and it is the last of the ten lighthouses built to protect ships along the Mississippi coast. Although the structure is approximately 320 feet from the shore today, it was originally at water's edge. The setback from the ocean is a result of the construction of the beach. Two important natural factors must be considered when any structure is built on the coast: geological setting and reoccurring hurricanes. The cast-iron shell has no doubt

The Biloxi Lighthouse, built on the shoreline in 1848, is now the only lighthouse in the United States to stand between the eastbound and westbound lanes of a highway. It survived the extraordinary storm surges of both Hurricanes Camille and Katrina.

played a role in the structure's survival, but it would not be here today without the considerable foundation work it received.

In 1860 a storm undermined the 1840s-era foundation, and the tower tilted 2 feet from vertical. During an 1866 renovation, sand was removed from the opposite side to straighten out the tower. The lighthouse has been renovated several times since then. The last renovation was completed in 2010. Markers on the brick walls inside indicate the storm surge levels of various hurricanes. The marks for Katrina and Camille remain the highest at 21.5 feet and 17.5 feet respectively.

Gulf Islands National Seashore

The exit for the Mississippi District headquarters of the Gulf Islands National Seashore is about 2 miles east of the junction of US 90 and MS 609, at Ocean Springs. The Mississippi District, which oversees five barrier islands and a mainland park, features natural beaches, historic sites, wildlife sanctuaries, bayous, nature trails, picnic areas, and campgrounds. However, the Davis Bayou Coastal Preserve in Ocean Springs is the only park accessible by automobile. Petit Bois, Horn, East Ship, West Ship, and Cat Islands are accessible only by boat. Ship Island is accessible by a ferry that leaves from Gulfport Harbor.

West Ship Island has the only natural deepwater harbor between the Mississippi River and Mobile Bay, in Alabama. Starting in the late eighteenth century the harbor provided anchorage for the ships of explorers, colonists, sailors, soldiers, defenders, and invaders. Fort Massachusetts was constructed on the island after the War of 1812. Early in the Civil War, Union forces took the half-finished fort from the Confederacy and utilized it for most of the war. In 1969 a survivor of Hurricane Camille rode out the storm and its surge in Fort Massachusetts, though the storm and its surge split the island in two. A similar feat would not have been possible during Hurricane Katrina, the storm surge of which topped the fort.

The islands offer excellent opportunities to observe many coastal processes at work, but one of the unexpected features of Mississippi's barrier islands is the accumulation of heavy mineral sand layers. The mineral grains were once part of the igneous and metamorphic rocks that formed the core of the Appalachian Mountains,

Dark, heavy mineral sand layers exposed in an eroded scarp on the Gulf side of Ship Island. Minerals such as ilmenite, rutile, and zircon are the remains of rocks weathered and eroded long ago from the Appalachian Mountains, brought to the Gulf by rivers, and redistributed by longshore drift. —Courtesy of Frank Heitmuller

and they have been reworked by rivers and the ocean over millions of years. Heavy mineral grains are more dense than quartz, the typical tan beach sand, and are also very resistant to erosion. Because they are relatively dense, wave activity separates them from less dense sand, and the heavy mineral sand accumulates in dark layers.

At least ten minerals have been found in the sands of Mississippi's barrier islands. The more abundant minerals include ilmenite, kyanite, staurolite, rutile, zircon, and magnetite. Ilmenite and rutile are titanium rich; zircon is rich in zirconium; and magnetite is iron rich. The heavy mineral sands on Mississippi's barrier islands have commercially exploitable titanium, but because the islands are part of the Gulf Islands National Seashore, it's unlikely the resources will ever be mined.

Belle Fontaine Drive, south of US 90, parallels Belle Fontaine Point. Wetlands of the Graveline Bay Coastal Preserve make up the eastern portion of the point. The wetlands developed on the Prairie Formation after the sea inundated the Graveline Creek valley during the Holocene. A wide section of the Gulfport Formation strand plain separates the preserve's wetlands from the coastline. Dark-brown, humate-impregnated (organic rich) Gulfport sands display hundreds upon hundreds of ghost shrimp (*Lepidophthalmus* species) tubes, or burrows, which are indicative of the shallow, nearshore environments that existed here during the Sangamon interglacial stage.

The Mississippi Sound is eroding the marsh deposits of Belle Fontaine Point. Along the beach the stumps of pine trees in growth position indicate the extent of erosion; the shoreline has retreated as much as 550 feet over the past 170 years. The purple- and brown-stained areas along the bank are where groundwater seeps out and the iron-bearing minerals it contains oxidize.

Belle Fontaine Drive to the west follows the full length of a 2.5-mile-long, narrow, low, dune-covered sand spit. The spit formed as longshore drift transported sand westward from the eroding Gulfport Formation bluffs of eastern Belle Fontaine Point. The spit and the entire subdivision built upon it were utterly devastated during Hurricane Katrina, as indicated by the numerous properties that still lie in ruins.

Hurricanes

Life on the Mississippi coast carries with it the potential for tropical storms and deadly hurricanes. Coastal populations continue to grow despite increased building restrictions and insurance rates. People are aware of the risks but are still drawn to the water. Storms and sea level rise are changing the coast, and people are forced to keep adapting to these changes. A drive along US 90 reveals how devastating the impacts of

Storm	Year	Landfall	Maximum Sustained Wind (mph)	Storm Surge (feet)
Camille	1969	Bay St. Louis/ Pass Christian	190	22.6
Elena	1985	Biloxi/Gulfport	125	6–8
Georges	1998	Biloxi	155	8.9
Katrina	2005	Pearlington	175	27.8

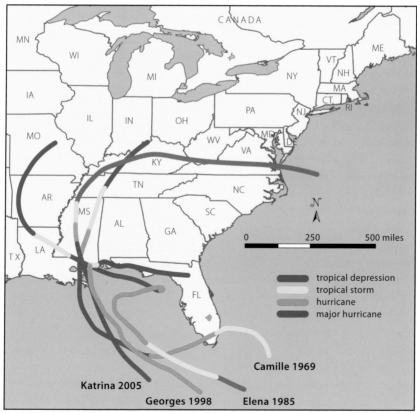

The paths of Mississippi's major hurricanes. (Based on data of the National Oceanic and Atmospheric Association.)

hurricanes can be to coastal residents. Ten years after Katrina, most residential lots along US 90 west of Gulfport have not been redeveloped.

Hurricanes are large, rotating, low-pressure storms known as tropical cyclones. In the northern hemisphere they rotate in a counterclockwise direction due to the effects of the Earth's rotation. Upward-moving air in the eye of the storm creates lower pressure than that encountered at the outer fringe of the storm. As a result, air is pulled inward toward the center of the storm, generating strong winds. When wind speed reaches 74 miles per hour, a tropical storm is considered a hurricane. In North America, the most intense period of storm activity is between mid-August and mid-September.

The Mississippi coast has been hit by several significant hurricanes and tropical storms in the last one hundred years. The eyes of Camille, Georges, Elena, and Katrina all made landfall near Biloxi. Many others (seventeen of them named) affected the coast, but these four were storms that left their mark. Tropical storms and hurricanes have brushed Biloxi fifty-two times in the last 143 years. Approximately 62 percent of them were of the less severe tropical-storm level. Hurricanes can be ranked by any number of factors, including damage, wind speed, diameter, air pressure, and storm surge. When measured by air pressure, two of the top-three strongest directly hit the Mississippi coast: Katrina at 920 millibars and Camille at 909 millibars. The 1935 Labor Day Hurricane in Florida ranks first at 892 millibars. Although Katrina ranks third in terms of air pressure, it still stands as the costliest natural disaster in US history at just over $1 billion. Approximately 33 percent of the insurance claims for this storm originated in Mississippi.

In addition to the strong winds and heavy rain associated with tropical storms and hurricanes are devastating storm surges. A storm surge is a rapid, abrupt rise in sea level above the predicted astronomical tide. It's created by strong wind blowing water landward. The greater the surge height, the farther inland the wave will carry. For example, debris was carried inland to I-10 during Hurricane Katrina.

Storm surges have had a significant effect on Mississippi's barrier islands as well as the coastline. The storm surges of significant hurricanes wash over Mississippi's barrier islands, changing their forms. Historic records indicate that Ship Island was frequently cut in half by storms. The storm surge of Hurricane Camille washed over Ship Island's dunes resulting in what is likely the permanent separation of West Ship and East Ship Islands. Islands with dune heights of less than 7 feet are most susceptible to breaching. Hurricane Katrina's storm surge washed away most of East Ship Island in 2005. Hurricanes Gustav and Ike nearly completely removed East Ship Island in 2008. Since that time, however, the island has reemerged.

Like so many natural phenomena, the height of storm surge depends on many factors, such as wind speed, air pressure, storm velocity, and storm size. For example, Hurricane Katrina landed as a category 3 storm with a 27.8-foot storm surge, yet with its record speed, Camille, a category 5 storm, only produced a 22.6-foot storm surge. Katrina's radius was double that of Camille's, which enabled it to mobilize four times the volume of water. Depending on the forward speed of hurricanes, storm

The high watermark at the level of the balcony of Beau Rivage Casino's
lighthouse records the maximum wave height ever recorded for an Atlantic
Basin storm. —Courtesy of Hermann Fritz, Georgia Institute of Technology

surges can be of short duration. Once the eye of the storm moves onto land the
surge dissipates. Though the greatest storm surge during Katrina was recorded at Pass
Christian (27.8 feet), a high watermark on the lighthouse at the Beau Rivage Casino, in
Biloxi, measured 34.1 feet above sea level. Here the surge only measured 22 feet, but
with a high tide and an 11.1-foot wave height, the addition of the storm surge created
the greatest high-water mark produced by an Atlantic hurricane.

Before Katrina in 2005, Hurricane Camille (1969) was the reference storm for
the Mississippi coast. There were few areas along the beachfront that Katrina's storm
surge didn't seriously damage or destroy. Most of the antebellum homes and other
buildings located on elevated beach ridges that avoided destruction during Camille
were destroyed. Even though they were designed to withstand hurricanes, many of
the large gambling barges moored along the Mississippi Sound were ripped from their
moorings and moved landward by Katrina's storm surge. As of 2015, many of the

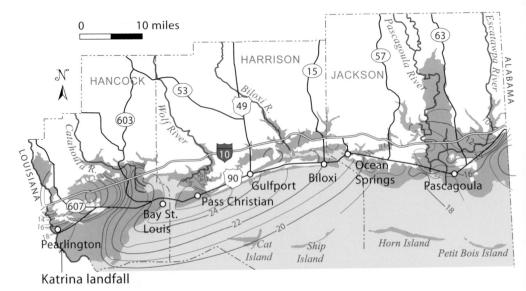

Katrina landfall

Hurricane Katrina made landfall near Pearlington on August 29, 2005. The gray areas represent land that was submerged by Katrina's storm surge, which reached 27.8 feet at Pass Christian. The contour lines represent the height of the storm surge in 1-foot increments. (Modified after Federal Emergency Management Agency 2005.)

One of the two casino barges that broke loose from moorings and were deposited on the north side of US 90 by Katrina's storm surge. The barge was 660 feet long and more than 125 feet wide. The storm surge moved it 1,800 feet to the northwest. —Courtesy of Lieutenant Commander Mark Moran, NOAA Corps, NMAO/AOC, https://imagespublicdomain.wordpress.com/category/noaa

buildings along Mississippi's coast, especially west of US 49 in Gulfport, have not been rebuilt. However, east of US 49 a lot more construction has taken place, especially in Biloxi and Ocean Springs. The gaming and hotel industries have been revitalized in Biloxi and D'Iberville as well.

From the eastern shore of Biloxi Bay to the Alabama state line, US 90 travels mainly on the Prairie Formation unless crossing floodplain alluvium of modern streams. The road crosses tidal marshland of the Pascagoula River for about 3 miles between Gautier and Pascagoula. Note the narrow, high, forested banks along some of the channels; these natural levees formed from sediment deposited during overbank flooding and high tides.

About 3.5 miles east of the junction with MS 63, US 90 crosses floodplain and marsh alluvium of the Escatawpa River. About 1 mile north of Orange Grove, the sizeable Escatawpa River makes a sudden 90-degree westward turn before discharging into the Pascagoula River. East of Moss Point, US 90 crosses the marshy relict of the Escatawpa channel that, in late Holocene time, nourished an active delta at Grand Bay, near the Alabama border. The Grand Bay National Wildlife Refuge occupies the marshy remnants of this delta. The Pascagoula captured the flow of the Escatawpa, which explains the dramatic bend. The abandoned river channel and marshy delta remnants are dotted with numerous Native American shell mounds (the equivalent of modern trash piles) that outline the former margins of the Escatawpa's channel. The middens suggest that the piracy took place in the late Holocene, sometime after human settlement of the region. Having long been deprived of river flow, the former delta is subsiding and steadily eroding.

MISSISSIPPI 63
US 90—US 98
41 miles
See the map on page 211.

Between the junction of US 90 and I-10, MS 63 traverses the Holocene-age floodplain of the Escatawpa River flanked by Pleistocene-age Prairie Formation deposits to the north and south. For about 8 miles north of I-10, MS 63 travels on the Prairie Formation, deposited by rivers on a relatively flat floodplain during the late Sangamon interglacial and the transition into the early Wisconsin glacial stage. Then, for about 10.5 miles (to about 0.5 mile north of Americus Road, or the Jackson-George county line) MS 63 travels on the western edge of the Montgomery Terrace, composed of silty sand, sand, and the occasional lignite bed. This former coastal plain developed between 221,000 and 176,000 years ago, during an interglacial stage that predates the Sangamon.

Between the Jackson-George county line and the US 98 junction at Lucedale, MS 63 mostly traverses sand, mud, clay, and occasional gravel of the Pliocene-age Citronelle Formation. The county line also marks where MS 63 passes from the Coastal Meadows physiographic province into the undulating topography of the Pine Hills physiographic province, which is underlain by the Citronelle.

Just north of the Jackson-George county line and south of Barton-Agricola Road there are numerous oval and circular depressions observable both on topographic maps and aerial images. These depressions, known as "blowouts," are up to 1,000 feet from the road and may be difficult to see, but many are ponds today. The larger depressions range up to but rarely over 500 feet in diameter; they are usually 5 to 15 feet deep but can be up to 40 feet deep. They formed as wind blew across a dry Citronelle surface, removing finer-grained sand in a process known as *deflation*. The studies of similar blowouts in Alabama determined that Alabama's blowouts have a northwest-southeast or north-south orientation, paralleling the wind direction at the time they formed. The orientation of the Mississippi blowouts hasn't been studied. The rounded outlines of many of the depressions suggest that after deflation they became water-filled ponds, and their shores were smoothed out and modified by wave action. The sand that was scoured from the blowouts would likely have formed dunes along their margins; their absence indicates the blowouts formed long ago, during the Pliocene, and water and wind have since carried away the dunes.

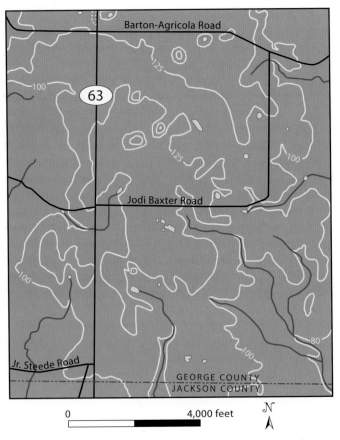

Blowouts are visible (some are now ponds) on this topographic map of southern George County. These depressions, on this relatively flat terrain, rarely exceed 500 feet in diameter and can be up to 40 feet deep.

About 3 miles north of the MS 26 junction the highway crosses Holocene-age floodplain and terrace alluvium of Big Creek. On either side of these deposits, at lower elevations, the road travels through small areas composed of clay and sand of the Miocene-age Hattiesburg and Pascagoula Formations.

MISSISSIPPI 605/ MISSISSIPPI 67
US 90—US 49
21 miles
See the map on page 211.

From the US 90 junction, Cowan Road (MS 605) traverses the Gulfport Formation strand plain. The subtle ridge and swale topography of the strand plain is noticeable for 0.8 mile between the railroad tracks and Pass Road. The topography here was not created by a meandering river channel, but rather by the sporadic shift in the beach ridge toward the south. Pass Road runs east-west along the apex of the highest Gulfport beach ridge. Bernard Bayou was a creek valley during Pleistocene time and delineates the northern edge of the Gulfport Formation. From the bayou to just north of I-10, MS 605 crosses over the Prairie Formation before entering the Citronelle surface, which it and MS 67 cross all the way to US 49.

The Prairie was deposited on a broad alluvial plain as sea level rose higher than its present position during the late Sangamon interglacial stage. The deposition of alluvium continued even after sea level dropped below its present elevation during the Wisconsin glaciation that followed. The Citronelle upland steadily rises northward. Gray, muddy, and clayey beds of the Miocene-age Hattiesburg and Pascagoula Formations underlie the usually silty and sandy, occasionally gravelly, yellowish-brown to brilliant orange Citronelle Formation.

GLOSSARY

absolute age. The age in years assigned to a rock based on isotopic dating methods.

abyssal. Depositional environment on the ocean floor at water depths that range between 10,000 and 20,000 feet.

accretion. A process that increases the size of a continent through the addition of smaller continental fragments.

acid. Solution that has a pH less than 7.0.

alkaline soil. Soil with a pH value that is greater than 7.0.

alluvial fan. Fan-shaped sedimentary deposit that forms when a stream becomes unconfined as it exits a region of higher topography.

alluvial plain. A level or gently sloping surface formed by stream-deposited sediments.˝

alluvium. Unconsolidated sediment deposited by rivers or streams.

anthracite. The highest grade of coal with a carbon content between 92 and 98 percent.

anticline. A concave-down geologic structure with folded rock layers that dip away from the central axis.

aquaculture. The farming of aquatic organisms.

aquifer. A porous body of sand, gravel, or rock that readily transmits water.

asthenosphere. The uppermost layer of the **mantle** that is not molten but has the ability to flow.

avulsion. The rapid abandonment of a **channel** and formation of a new channel along another **course.**

axis. A line delineating the maximum curvature in folded rocks.

backswamp. The area behind natural **levees** where **clay** and **silt** are deposited during flooding.

backwater flooding. Flooding that occurs when water levels rise high enough in the main **channel** to inhibit the flow of tributary streams.

barrier island. A long and narrow offshore sandbar that rises out of the water parallel to the coastline and is separated from the mainland by a **lagoon.**

basin. A depressed area that collects sediment, or a structural fold in which all rock layers dip toward the center of the feature.

bayou. A shallow **wetland** lake or river generally found in low-lying areas with slow-moving currents

beach nourishment. A process in which **sand** is purposefully pumped from offshore areas to form and maintain artificial beaches.

beach ridge. A wave- or wind-deposited ridge of **sand** that forms parallel to and landward of the beach.

bed. A layer of rock that is distinct from beds below and above it.

bed load. Sediment that rolls, slides, or bounces along the bottom of a river **channel**.

bedrock. Solid rock that lies beneath unconsolidated material.

bluff. A high, steep cliff.

bluff line. A broad, rounded cliff bordering a **floodplain**, river, or beach.

boulder. A sedimentary particle with an average diameter greater than 10 inches (256 mm).

brachiopod. A marine bivalve with bilateral symmetry and two shells of unequal size.

braided stream. A river or stream that features multiple crossing channels and numerous sandbars or gravel bars. Indicative of highly variable flow rates and sediment loads.

bryozoan. A small colonial organism that builds calcareous skeletons similar to corals.

buhrstone. A **silica**-rich rock used for grinding grain.

calcite. A common mineral composed of **calcium carbonate**. The chief component of **limestone**.

calcium carbonate. A chemical compound composed of calcium, carbon, and oxygen.

calving. A process in which large blocks of **loess** break away from exposed faces along vertical fractures.

carbonate. A sedimentary rock composed of **limestone** or **dolomite**.

carbonate bank. A **depositional environment** located in tropical latitudes with clear, shallow water.

cast. A type of **fossil** created when a mineral or sediment fills a void in sediment or rock left behind by an organism.

cementation. The growth of new minerals around sedimentary grains. Common cements are **quartz**, **calcite**, and iron oxide.

chalk. A **limestone** composed mainly of the shells of one-celled planktonic algae called **coccolithophorids**.

channel. The deeper part of a stream or river.

channelization. The process of straightening or shortening a **channel** to enable water to flow faster and to reduce flooding.

chemical weathering. A process that decomposes rocks into various components that are different from the rock's original composition.

chert. A common **mineral** composed of silicon and oxygen with the same composition as quartz but featuring a microcrystalline structure.

clast. A fragment of a preexisting rock.

clastic. A sedimentary rock composed of fragments of preexisting rocks that have been transported from their original locations.

clay. A mineral formed by **chemical weathering**. Also, a grain with an average diameter less than 0.0625 millimeter.

coastal plain. A low-relief, low-elevation plain adjacent to the ocean.

cobble. A sedimentary particle with an average diameter between 2.5 inches (64 mm) and 10 inches (256 mm).

coccolith. Oval plates that compose the shell of a **coccolithophorid**. A primary component of **chalk**.

coccolithophorid. One-celled planktonic algae that secrete a shell composed of **calcium carbonate**.

concretion. A nodular or spherical accumulation of grains or precipitated minerals.

contact. The surface where two different types of rocks come together.

continental shelf. Gently sloping plains adjacent to coastlines and covered by shallow water. Water depth may range up to 500 feet.

course. An abandoned **channel** of a river that is longer than an **oxbow** and traceable for several miles.

crevasse. A break in a **levee** or riverbank.

crossbedding. Secondary bedding in a sedimentary rock that is at an angle to the main bedding plane.

cross section. A profile of a vertical section through the Earth.

crust. The outermost layer of the Earth that is divided into continental crust and oceanic crust.

cuesta. A landform with a gentle slope on one side and steep slope on the other; created by a dipping rock unit.

cut bank. The steep bank located on the outer edge of a river or stream **meander**.

daughter isotope. An element produced by the **radioactive decay** of an unstable, or parent, isotope.

delta. A body of sediment that forms at the mouth of a river.

Delta. The diamond-shaped region that lies between Memphis, Tennessee, and Vicksburg, Mississippi, and is east of the Mississippi River and west of the **bluff line**.

depositional environment. The condition and location in which sediment is deposited. Environments are classified as **marine, terrestrial,** or transitional.

differential erosion. Softer materials, such as **clay** or shale, or weaker materials, such as loose **sand** or fractured rock, that erode more easily than those that are better cemented or are not fractured.

dip. The angle measured from the horizontal plane that is used to describe the orientation of sedimentary rock layers or **fault planes**.

discharge. The volume of water that passes a point per unit time.

dissolved load. The portion of **stream load** that is dissolved in solution.

distributary. An outflowing branch of a river **channel** that forms in a **crevasse** and may or may not rejoin the main channel.

dolomite. A light-colored mineral composed of calcium magnesium carbonate.

dome. A structural fold in which all rock layers dip away from the center of the feature.

drainage basin. The total surface area that contributes runoff to a river or stream.

dune. A windblown mound of loose sediment generally found in deserts or coastal areas.

erosion. The movement or transport of weathered material by wind, water, gravity, or ice.

escarpment. A steep slope at the edge of a plateau.

estuary. The portion of a river near its mouth that is influenced by tides.

fault. A fracture in the Earth along which there has been relative displacement of material on either side of the fracture and the motion was parallel to the fracture.

fault plane. The nearly planar surface along which displacement takes place along a **fault**.

floodplain. The area adjacent to a river **channel** that is flooded when the water rises above **flood stage**.

flood stage. The elevation a river reaches when it overflows its banks or **levees** and flows onto the **floodplain**.

fold. Describes rocks that have been bent by compressive **tectonic** stress.

footwall. The side of a **fault** that is beneath the **fault plane**.

formation. A unit of rock that is distinctive and can be identified and mapped over a large area.

fossils. The remains, traces, or tracks of plants and animals that have been preserved.

gastropod. An organism, such as a snail, that secretes an asymmetrically coiled, one-chambered, calcium carbonate shell.

geode. A generally hollow, spherical rock lined with crystals that **precipitated** out in the void space.

glacial stage. A period during an ice age when global temperatures are lowest and ice is prevalent.

glaciation. A period of time during which continental glaciers are extensive.

glauconite. A green, iron-rich mineral that is a common component in sedimentary rocks of marine origin.

graben. A linear valley that forms when the ground between two **normal faults** drops down relative to the ground on the other side of the faults.

gradient. The change in elevation between two points along a river divided by the horizontal length of the **channel**.

gravel. Sediment with an average diameter between 2 and 64 millimeters (2.5 inches).

group. A mapping unit composed of two or more **formations**.

gypsum. A common **mineral** made of calcium, sulfur, and oxygen. The primary component in plaster of paris.

halite. A common **mineral** made of sodium and chlorine. Salt.

hanging wall. The side of a **fault** that is above the **fault plane**.

hot spot. A volcanic province above an area in the **mantle** where abnormally high heat escapes from deep within the mantle.

hurricane. A large, tropical cyclonic storm with minimum wind speeds of 74 miles per hour.

hydrocarbon. An organic compound made of hydrogen and carbon. Examples include oil and natural gas.

hydrophyte. A plant adapted to growing in water-saturated soils.

inner core. The innermost part of the Earth that is composed of solid iron and nickel.

interbedded. When two or more types of sedimentary rock are alternately layered.

interglacial stage. A period during an ice age when global temperatures are similar to today's temperatures.

joint. A fracture in rock along which there has been no displacement.

knickpoint. A point in a stream where there are sudden changes in **gradient**, such as at waterfalls and rapids.

lag deposit. An accumulation of coarse particles that is left after water or wind removes the finer particles.

lagoon. A shallow body of water separated from a larger body of water by a **barrier island** or reef.

lava. Molten rock on the surface of the Earth.

levee. A natural or engineered linear ridge that parallels a **channel** and limits flooding. Natural levees form when floodwater drops sand and silt that can no longer be carried in suspension.

lignite. Soft coal that contains traces of plant matter and has a carbon content between 46 and 60 percent.

limestone. A sedimentary rock composed of the mineral **calcite**.

lithosphere. The rigid, outermost layer of Earth that includes the **crust** and upper **mantle**, to a depth of approximately 60 miles.

loess. Nonstratified, windblown silt of glacial origin.

magma. Molten rock beneath Earth's surface.

magnetic reversal. The reversal of Earth's magnetic field. Earth's magnetic poles reverse, on average, every 200,000 years.

mantle. The largest layer of Earth, between the **outer core** and **crust**, that is composed of iron and magnesium silicate.

marine. Oceanic **depositional environments**.

marl. A general term for clay-rich or sandy **limestone** that is 30 to 90 percent **carbonate**.

marsh. A **wetland** characterized by **hydrophytes** and water-saturated soils.

matrix. The fine-grained material that encases larger sediments (such as **cobbles**), **fossils**, or mineral crystals.

meander. A large, pronounced bend in a river **channel**.

member. A rock unit that is distinct from the **formation** it is associated with but cannot be mapped over a large area.

mica. A general term for a group of **silica**-rich minerals that tend to form thin, flexible sheets.

mid-ocean ridge. A linear volcanic ridge of oceanic **crust** that forms from **magma** upwelling from the **mantle**.

mineral. A naturally occurring solid substance with a characteristic crystal form.

mold. The internal or external impression of a **fossil** or **trace fossil**.

mud. A combination of clay- and silt-sized grains with diameters less than 0.0625 millimeters.

normal fault. A break in the Earth's **crust**, caused by **tensional stress**, along which the **hanging wall** block drops down relative to the **footwall** block.

nourishment. See **beach nourishment**.

orthoquartzite. A sandstone that is so strongly cemented with **silica** that it breaks across the sand grains.

outcrop. Bedrock exposed at the surface.

outer core. The layer of Earth that is composed of molten iron and nickel and lies between the **inner core** and the **mantle**.

overbank. The area outside of a river's **levee**.

oxbow. An arcuate lake that is created when a river **meander** is cut off and that segment of river is abandoned.

parent isotope. An unstable element that decays to produce **daughter isotopes**.

passive continental margin. A continental plate boundary free of **tectonic** activity.

permeable. An Earth material with pores that are interconnected and permit the flow of gas or liquids.

physiographic province. A geographic region with similar topography, **bedrock**, and often structural geology that result in characteristic relief and ecology.

piercement dome. A **salt dome** that breaks through overlying layers of rock and sediment as it rises toward the surface.

plankton. Very small organisms that float in oceans, lakes, or rivers.

plate tectonics. A theory that posits that Earth's **lithosphere** is broken into **tectonic** plates that move and interact with each other, leading to landforms such as mountains and ocean **basins**.

point bar. A deposit of **sand** or **gravel** deposited on the inner bank of a **meander** loop.

pore. Void space in a rock.

precipitate. A mineral that forms when a solution evaporates. To be separated from a solution by a chemical process.

pre-loess. Sediment deposited prior to the windblown **loess** deposited in the Pleistocene Epoch.

pyrite. A mineral composed of iron sulfide; also known as fool's gold.

quartz. A mineral composed of silicon dioxide, also referred to as **silica**.

quartzite. A metamorphic rock formed from sandstone that was exposed to elevated temperature and pressure.

radioactive decay. A process in which the nucleus of an unstable element loses energy in the form of alpha particles, beta particles, or gamma radiation.

radioisotope. A radioactive element.

radiometric dating. A method of **absolute** age dating that determines the age of a sample by determining the relative proportions of particular **radioisotopes**.

recharge zone. An area of exposed Earth materials that enables water to infiltrate an **aquifer**.

regression. The movement of terrestrial **depositional environments** toward the ocean through time.

regressive. A sequence of rocks suggesting a lowering of sea level or a rise in the land surface.

relative age. The age of rock units relative to one another in sequence, not years.

relief. The difference between the greatest and least elevations in an area.

replacement. The substitution of either elements or cellular material by other elements introduced by flowing groundwater.

reservoir rock. Porous and **permeable** rock that contains oil or gas.

reverse fault. A break in the Earth's crust caused by compression along which the **hanging wall** block rises relative to the **footwall** block.

ridge and swale. A series of parallel ridges created by the advance of a **point bar** across a **floodplain**.

rift. A large-scale linear valley created by **tensional stress**.

salt dome. A structural feature caused by the upward movement of salt due to its low density.

sand. A grain with an average diameter between 0.0625 and 2 millimeters.

silica. The compound silicon dioxide.

silt. A grain with an average diameter between 0.0039 and 0.0625 millimeter.

slump. A type of slope failure in which the mass of rock and sediment moves downward and rotates along a scoop-shaped fracture.

source rock. A rock unit rich in organic carbon that generates hydrocarbons with burial and subsequent heating.

splay. A sloping and spreading sand deposit created when a **crevasse** develops in a **levee**.

spring. A place where water naturally flows from rock or sediment.

storm surge. The wind-driven increase in coastal water levels related to an approaching large storm, such as a tropical storm or **hurricane**.

strand plain. A broad belt of **sand** along a shoreline featuring **ridges and swales**.

strata. Multiple layers of sedimentary rock.

stream load. Material carried by water in a stream. Consists of the **bed load**, **suspended load**, and **dissolved load**.

stream piracy. When water in a stream or river is diverted into a **channel** of an adjacent river.

strike-slip fault. A break in the Earth's crust along which the displacement is horizontal rather than vertical.

subduction zone. Where an oceanic plate descends beneath a continental plate and dives into the **asthenosphere**.

suspended load. The portion of total **stream load** that is transported in the moving water.

syncline. A convex-down geologic structure with folded rock layers that dip toward the central **axis**.

tectonic. Relates to the deformation of rocks related to **plate tectonic** activity.

tensional stress. Directed stress that forces material away from a central imaginary plane.

terrace. A level, flat area above the current **floodplain** that represents a former floodplain surface.

terrace deposit. **Sand** and **gravel** that accumulated as a river meandered across a former **floodplain**. Often found at elevations much higher than any modern stream.

terrestrial. A **depositional environment** located on land.

thrust fault. A **reverse fault** with a low-angle **fault plane**.

tidal marsh. A **marsh** in coastal areas that is periodically inundated with water as tides rise and fall.

trace fossil. Indirect evidence of life, such as tracks, footprints, trails, burrows, and feces.

transform boundary. A **plate tectonic** boundary where plates slide horizontally past each other.

transform fault. A **fault** that connects offset portions of a **mid-ocean ridge**.

transgression. The movement of marine **depositional environments** toward land through time.

transgressive. A sequence of rocks suggesting a rise in sea level or a lowering of the land surface.

trench. A linear, very deep depression on the seafloor located above a **subduction zone**.

tributary. A stream that discharges into another stream.

trilobite. A marine arthropod that thrived during the Paleozoic Era.

tsunami. A large wave created by a subsea earthquake, landslide, or meteorite impact.

type section. The outcrop where a stratigraphic unit is first described.

unconformity. A surface separating strata of significantly different ages that represents a period of erosion; a gap in geologic time.

watershed. The area that contributes runoff to a river.

weathering. The decomposition of rocks through chemical processes or the disintegration of rocks through physical processes.

wetland. A geographic area featuring **hydrophytes**, soils that reflect characteristics of being submerged by water, and high water tables for part of the growing season.

yazoo stream. A tributary that flows parallel to a river for a long distance before joining the river.

REFERENCES

Allmon, W. D., J. A. Picconi, S. F. Greb, and C. C. Smith. 2011. Geologic History of the Southeastern U.S.: Reconstructing the Geologic Past. http://geology .teacherfriendlyguide.org/downloads/se/tfggse_1_geolhistory.pdf.

American Shore and Beach Preservation Association. 2007. *Beach Nourishment: How Beach Nourishment Projects Work*. Available at http://www.asbpa.org /publications/fact_sheets/HowBeachNourishmentWorksPrimerASBPA.pdf.

Anderson, D. G., T. G. Bissett, and S. J. Yerka. 2013. "The Late Pleistocene Human Settlement of Interior North America: The Role of Physiography and Sea Level Change." In *Paleoamerican Odyssey*, edited by K. E. Graf, C. V. Ketron, and M. R. Waters, 235–255. College Station: Texas A & M University Press.

Arthur, J. K., and R. E. Taylor. 1998. *Ground-Water Flow Analysis of the Mississippi Embayment Aquifer System, South-Central United States*. US Geological Survey Professional Paper 1416-I.

Autin, W. J. 1996. "Pleistocene Stratigraphy in the Southern Lower Mississippi Valley." *Engineering Geology* 45 (1–4): 87–112.

Bird, D. E., K. Burke, S. A. Hall, and J. F. Casey. 2005. "Gulf of Mexico Tectonic History: Hotspot Tracks, Crustal Boundaries, and Early Salt Distribution." *American Association of Petroleum Geologists Bulletin* 89 (3): 311–328.

Bird, D. E., K. Burke, S. A. Hall, and J. F. Casey. 2011. "Tectonic Evolution of the Gulf of Mexico Basin." In *Gulf of Mexico: Origin, Waters, and Biota*, vol. 3, Geology. Edited by N. A. Buster and C. W. Holmes, 3–16. College Station: Texas A & M University Press.

Blain, W. T. 1976. *Education in the Old Southwest: A History of Jefferson College, Washington, Mississippi*. Washington, MS: Friends of Jefferson College.

Blake, E. S., and E. J. Gibney. 2011. *The Deadliest, Costliest, and Most Intense United States Tropical Cyclones from 1851 to 2010 (and Other Frequently Requested Hurricane Facts)*. National Oceanic and Atmospheric Administration Technical Memorandum NWS NHC-6. Available at http://www.nhc. noaa.gov/pdf/nws-nhc-6.pdf.

Blaustein, R. J. 2001. "Kudzu's Invasion into Southern United States Life and Culture." In *The Great Reshuffling: Human Dimensions of Invasive Alien Species*, edited by J. A. McNeely, 55–62. Gland, Switzerland; Cambridge, UK: IUCN.

Blum, M. D., M. J. Guccione, D. A. Wysocki, P. C. Robnett, and E. M. Rutledge. 2000. "Late Pleistocene Evolution of the Lower Mississippi River Valley, Southern Missouri to Arkansas." *Geological Society of America Bulletin* 112 (2): 221–235.

Bograd, M. 2014. *Earthquake Epicenters in Mississippi*. Mississippi Department of Environmental Quality Fact Sheet 1.

Borcherdt, R. D., J. H. Healy, W. H. Jackson, and D. H. Warren. 1967. *Seismic Measurements of Explosions in the Tatum Salt Dome, Mississippi*. US Geological Survey Open-File Report 67–24.

Brown, B. W. 1960. *Geologic Study Along Highway 16 From Alabama Line to Canton, Mississippi*. Mississippi Geological Survey Bulletin 89. Available at http://deq.ms.gov/MDEQ.nsf/page/Geology_Bulletin89GeologicStudy AlongHighway16FromAlabamaLinetoCantonMississippi?OpenDocument.

Burgess, W. J. 1976. "Geologic Evolution of the Mid-Continent and Gulf Coast Areas: A Plate Tectonics View." In *Gulf Coast Association of Geological Societies Transactions*, vol. 26, 132–143.

Cathcart, T., and P. Mellby. 2009. *Landscape Management and Native Plantings to Preserve the Beach Between Biloxi and Pass Christian, Mississippi*. Office of Agricultural Communications Bulletin 1183, Mississippi State University. Available at http://mafes.msstate.edu/publications/bulletins/b1183.pdf.

Clark, B. R., M. H. Rheannon, and J. J. Gurdak. 2011. *Groundwater Availability of the Mississippi Embayment*. US Geological Survey Professional Paper 1785.

Cox, R. T., and R. B. Van Arsdale. 1997. "Hotspot Origin of the Mississippi Embayment and Its Possible Impact on Contemporary Seismicity." *Engineering Geology* 46 (3–4): 201–216.

Cuevas, J. 2011. *Cat Island: The History of a Mississippi Gulf Coast Barrier Island. Jefferson*, NC: McFarland and Co.

Danehy, D. R., P. Wilf, and S. A. Little. 2007. "Early Eocene Macroflora from the Red Hot Truck Stop locality (Meridian, Mississippi, USA)." *Palaeontologia Electronica* 10 (3): 17A. Available at http://palaeo-electronica. org/2007_3/132/132.pdf.

Dockery, D. T., III. 1977. *Mollusca of the Moodys Branch Formation, Mississippi*. Mississippi Geologic, Economic, and Topographic Survey Bulletin 120. Available at https://www.deq.state.ms.us/mdeq.nsf/page/Geology_Mollusca oftheMoodysBranchFormationMississippi?OpenDocument.

Dockery, D. T., III. 1986. "Punctuated Succession of Paleogene Mollusks in the Northern Gulf Coastal Plain." *Palaios* 1 (6): 582–589.

Dockery, D. T., III. 1996. "Toward a Revision of the Generalized Stratigraphic Column of Mississippi." *Mississippi Geology* 17 (1): 1–8.

Dockery, D. T., III. 1997. *Windows into Mississippi's Geologic Past*. Mississippi Department of Environmental Quality, Office of Geology Circular 6. Available at https://www.deq.state.ms.us/mdeq.nsf/page/Geology_Circular-Windows IntoMississippi'sGeologicalPast?OpenDocument.

Dockery, D. T., III. 1998. "Molluscan Faunas Across the Paleocene/Eocene Series Boundary in the North American Gulf Coastal Plain." In *Late Paleocene-Early Eocene Biotic and Climatic Events in the Marine and Terrestrial Records.* Edited by M. P. Aubry, S. G. Lucas, and W. A. Berggren, 296–322. New York: Columbia University Press.

Dockery, D. T., III. 2008a. *Cenozoic Stratigraphic Units in Mississippi.* Mississippi Office of Geology. Available athttps://www.deq.state.ms.us/mdeq.nsf /pdf/Geology_MSCenozoicStratigraphy/$File/MS_Cenozoic_Stratigraphy .pdf?OpenElement.

Dockery, D. T., III. 2008b. *Mesozoic Stratigraphic Units in Mississippi.* Mississippi Office of Geology. Available at https://www.deq.state.ms.us/mdeq.nsf /pdf/Geology_MSMesozoic Stratigraphy/$File/MS_Mesozoic_Stratigraphy .pdf?OpenElement.

Dockery, D. T., III. 2008c. *Paleozoic Stratigraphic Units in Mississippi.* Mississippi Office of Geology. Available at https://www.deq.state.ms.us/mdeq.nsf /pdf/Geology_MSPaleozoicStratigraphy/$File/MS_Paleozoic_Stratigraphy .pdf?OpenElement.

Dockery, D. T., III, J. C. Marble, and J. Henderson. 1997. "The Jackson Volcano." *Mississippi Geology* 18 (3): 33–45. Available at https://www.deq.state .ms.us/mdeq.nsf/pdf/Geology_Volume018Number3September1997/$File /Vol_18_3.pdf?OpenElement.

Federal Emergency Management Agency. 2005. *Mississippi Hurricane Katrina Surge Inundation and Advisory Base Flood Elevation Maps: Harrison County, Mississippi.* Available athttps://www.fema.gov/response-recovery/hurricane-katrina-surge-inundation-and-advisory-base-flood-elevation-maps-1.

Flocks, J., and E. Klipp. 2009. "Storm Impact, Sea-Floor Change, and Barrier-Island Evolution: Scientists Map the Sea Floor and Stratigraphy around Ship and Horn Islands, Northern Gulf of Mexico." *Soundwaves* 113. Available at http://soundwaves.usgs.gov/2009/03/.

Foster, V. M., and T. E. McCutcheon. 1941. *Forrest County Mineral Resources.* Mississippi State Geological Survey Bulletin 44. Available at https://www .deq.state.ms.us/MDEQ.nsf/page/Geology_Bulletin44FORRESTCOUNTY MINERALRESOURCES?OpenDocument.

Galloway, W. E., P. E. Ganey-Curry, X. Li, and R. T. Buffler. 2000. "Cenozoic Depositional History of the Gulf of Mexico Basin." *American Association of Petroleum Geologists Bulletin* 84 (11): 1743–1774.

Galloway, W. E. 2011. "Pre-Holocene Geological Evolution of the Northern Gulf of Mexico Basin." In *Gulf of Mexico Origin, Waters, and Biota*, vol. 3, Geology. Edited by N. A. Buster and C. W. Holmes, 33–52. College Station: Texas A & M University Press.

Geological Society of America. 2012. *Geologic Time Scale*. Available at http://www.geosociety.org/science/timescale/.

Gilliland, W. A., and D. W. Harrelson. 1978. *Clarke County Geology and Mineral Resources*. Mississippi Department of Natural Resources, Bureau of Geology Bulletin 121. Available at https://www.deq.state.ms.us/mdeq.nsf/page/Geology_Bulletin121?OpenDocument.

Grand Bay National Estuarine Research Reserve. 2013. *Grand Bay National Estuarine Research Reserve Management Plan 2013–2018*. Moss Point, MS: Grand Bay National Estuarine Research Reserve, Mississippi Department of Marine Resources. Available at https://coast.noaa.gov/data/docs/nerrs/Reserves_GRD_MgmtPlan.pdf.

Green, S. R. 1985. "An Overview of the Tennessee-Tombigbee Waterway." *Environmental Geology and Water Science* 7 (1): 9–13.

Haq, B. U., J. Hardenbol, and P. R. Vail. 1987. "Chronology of Fluctuating Sea Levels Since the Triassic." *Science* 235: 1156–1167.

Harrison County. 2015. Sand Beach Plan Appendix: Background Assessment. Available at http://co.harrison.ms.us/downloads/departmental%20downloads/sand%20beach/Sand%20Beach%20Plan-Background.pdf.

Harry, D. L., J. Londono, and A. Huerta. 2003. "Early Paleozoic Transform-Margin Structure Beneath the Mississippi Coastal Plain, Southeast United States." *Geology* 31 (11): 969–972.

Hart, R. M., B. R. Clark, and S. E. Bolyard. 2008. *Digital Surfaces and Thicknesses of Selected Hydrogeologic Units Within the Mississippi Embayment Regional Aquifer Study (MERAS)*. US Geological Survey Scientific Investigations Report 2008–5098.

Hooke, J. M. 2004. "Cutoffs Galore!: Occurrence and Causes of Multiple Cutoffs on a Meandering River." *Geomorphology* 61 (3–4): 225–238.

Hoseman, D. 2015. *Cat Island Fact Sheet*. Available at https://www.sos.ms.gov/Education-Publications/Documents/Downloads/CatIslandPamphlet.pdf.

Huber, M. S., D. T. King, L. W. Petruny, and C. Koeberl. 2013. "Revisiting Kilmichael (Mississippi), A Possible Impact Structure." In *44th Lunar and Planetary Science Conference*, Contribution no. 1719, 2250.

Illinois South Tourism Bureau. 2015. *Cahokia Mounds: Collinsville, Illinois*. Available at http://cahokiamounds.org/wpress/wp-contentuploads/2015/09/Site-Brochure-8-13.pdf.

Ingram, S. L. 1991. "The Tuscahoma-Bashi Section at Meridian, Mississippi: First Notice of Lowstand Deposits above the Paleocene-Eocene TP2/TE1 Sequence Boundary." *Mississippi Geology* 11 (4): 9–14.

Ireland, H. A. 1974. "Query: Orthoquartzite????" *Journal of Sedimentary Petrology* 44 (1): 264–265.

Jennings, S. P. 2001. *Preliminary Report on the Hydrogeology of the Tertiary Aquifers in the Alluvial Plain Region of Northwestern Mississippi—Subcrop Geologic Map and Cross Sections.* Mississippi Department of Environmental Quality, Office of Land and Water Resources Open-File Report 01–101.

Keady, D. M. 1962. *Geologic Study along Highway 45 from Tennessee Line to Meridian, Mississippi.* Mississippi Geological Survey Bulletin 94. Available at http://deq.ms.gov/MDEQ.nsf/page/Geology_Bulletin94GeologicStudyAlongHighway45FromTennesseeLinetoMeridianMississippi?OpenDocument.

KellerLynn, K. 2010. *Geological Resources Inventory Scoping Summary, Natchez Trace Parkway: Mississippi, Alabama, Tennessee.* Geological Resources Division, National Park Service. Available at http://www.nature.nps.gov/geology/inventory/publications/reports/NATR_GRI_scoping_summary_2010-0804.pdf.

Kidd, J. T. 1975. "Pre-Mississippian Subsurface Stratigraphy of the Black Warrior Basin in Alabama." In *Gulf Coast Association of Geological Societies Transactions*, vol. 25, 20–39.

Kirgan, H. 2011. "Proposed Richton Salt Dome Project Is Dead as Funding for Environmental Impact Statement Cancelled." Gulflive.com (blog). http://blog.gulflive.com/mississippi-press-news/2011/09/proposed_richton_salt_dome_pro.html.

Knabb, R. D., R. Rhome, and D. P. Brown. 2011. *Tropical Cyclone Report, Hurricane Katrina, 23–30 August 2005.* National Hurricane Center. Available at http://www.nhc.noaa.gov/data/tcr/AL122005_Katrina.pdf.

Koeberl, C., W. U. Reimold, D. T. King Jr., and S. L. Ingram Sr. 2000. "The Kilmichael Structure, Mississippi: No Evidence for an Impact Origin from a Preliminary Petrographic Study." *Lunar and Planetary Science* 31: 1602.

Kolb, C. R., W. B. Steinriede Jr., E. L. Krinitzsky, R. T. Saucier, P. R. Mabrey, F. L. Smith, and A. R. Fleetwood. 1968. *Geological Investigation of the Yazoo Basin, Lower Miss. Valley.* US Army Engineer Waterways Experiment Station, Corps of Engineers Technical Report 3–480.

Kominz, M. A. 2001. "Sea Level Variations over Geologic Time." In *Encyclopedia of Ocean Sciences*, edited by J. H. Steele, S. A. Thorpe, and K. K. Turekian, 2605–2613. San Diego: Academic Press.

Lambeck, K., H. Rouby, A. Purcell, Y. Sun, and M. Sambridge. 2014. "Sea Level and Global Ice Volumes from the Last Glacial Maximum to the Holocene." *Proceedings of the Academy of Sciences* 111 (43): 15296–15303.

Land, L. S., J. A. Kupecz, and L. E. Mack. 1988. "Louann Salt Geochemistry (Gulf of Mexico Sedimentary Basin, U.S.A.): A Preliminary Synthesis." *Chemical Geology* 74 (1–2): 25–35.

Lighthousefriends.com. n.d. http://www.lighthousefriends.com/light.asp?ID=543.

Logan, W. N. 1908. *Clays of Mississippi: Part II.* Mississippi State Geological Survey Bulletin 4. Available at https://www.deq.state.ms.us/MDEQ.nsf /page/Geology_Bulletin4CLAYSOFMISSISSIPPI,PARTII,BrickClaysand ClayIndustryofSouthernMississippi?OpenDocument.

Lord, A. S., C. A. Rautman, and K. M. Looff. 2007. *Geologic Technical Assessment of the Richton Salt Dome, Mississippi, for Potential Expansion of the U.S. Strategic Petroleum Reserve.* Sandia National Laboratories Report SAND2007-0463. Available at http://prod.sandia.gov/techlib/access-control .cgi/2007/070463.pdf.

Lowe, E. N. 1913. *Preliminary Report on Iron Ores of Mississippi.* Mississippi State Geological Survey Bulletin 10. Available at https://www.deq.state .ms.us/mdeq.nsf/page/Geology_Bulletin10PremininaryReportonIronOres ofMississippi?OpenDocument.

Lusk, T. W. 1963. "Problem of Desiccation Sinking at Clarksdale." In *Mississippi Geologic Research Papers-1962*, Mississippi Geological, Economic, and Topographical Survey Bulletin 97, edited by M. K. Kern, W. H. Moore, T. W. Lusk, L. Hubricht, and E. H. Rainwater, 60–76. Available at https://www .deq.state.ms.us/mdeq.nsf/pdf/Geology_Bulletin97/$File/Bulletin%2097 .pdf?OpenElement.

Lüthi, D., M. Le Floch, B. Bereiter, T. Blunier, J. M. Barnola, U. Siegenthaler, D. Raynaud, et al. 2008. "High Resolution Carbon Dioxide Concentration Record 650,000–800,000 Years Before Present." *Nature* 453: 379–382.

Mann, C. J., and W. A. Thomas. 1968. "The Ancient Mississippi River." In *Gulf Coast Association of Geological Societies Transactions*, vol. 18, 187–204.

Marshall, C. 2014. "Can One Power Plant Clean Up Coal and Make Money?" *Scientific American* (July 23). http://www.scientificamerican.com/article /can-1-power-plant-clean-up-coal-and-make-money/.

Masters, J. n.d. "A Detailed View of the Storm Surge: Comparing Katrina to Camille." Weather Underground. http://www.wunderground.com/hurricane/surge_details.asp?

Matson, G. C. 1917. "The Catahoula Sandstone." In *Shorter Contributions to General Geology, 1916*, US Geological Survey Professional Paper 98-M, edited by D. White, 209–226. Available at http://pubs.usgs.gov/pp/0098m /report.pdf.

Merrill, R. K. 1983. "Source of the Volcanic Precursor to the Upper Cretaceous Bentonite in Monroe County, Mississippi." *Mississippi Geology* 3 (4): 1–6.

Merrill, R. K., D. E. Gann, and S. P. Jennings. 1988. *Tishomingo County Geology and Mineral Resources.* Mississippi Bureau of Geology Bulletin 127. Available at http://www.deq.state.ms.us/mdeq.nsf/page/Geology_Bulletin 127TishomingoCountyGeologyandMineralResources.

Merrill, R. K., J. J. Sims Jr., D. E. Gann, and K. J. Liles. 1985. *Newton County Geology and Mineral Resources.* Mississippi Department of Natural Resources, Bureau of Geology Bulletin 126. Available at http://www.deq.state.ms.us

/mdeq.nsf/page/Geology_Bulletin126NEWTONCOUNTYGEOLOGYAN
DMINERALRESOURCES?OpenDocument.

Meyer-Arendt, K. J. 2005. "Beach and Nearshore Sediment Budget of Harrison County, Mississippi: A Historical Analysis." In *Mississippi Coastal Geology and Regional Marine Study 1990–1994*, vol. 4, 437–489. St. Petersburg, FL: US Geological Survey Center for Coastal Geology and Regional Marine Studies.

Mills, H. H., and J. M. Kaye. 2001. "Drainage History of the Tennessee River: Review and New Metamorphic Quartz Gravel Locations." *Southeastern Geology* 40 (2): 75–97.

Mississippi Automated Resource Information System. 2008. Statewide 10-Meter Digital Elevation Model. http://www.maris.state.ms.us/HTM/Download Data/DEM.html.

Mississippi Casinos.com. 2016. "Tour." http://www.mississippicasinos.com/TourPostKatrina.htm.

Mississippi Department of Environmental Quality. 2015. Geologic Map of Mississippi Shape Files. Office of Geology. http://www.deq.state.ms.us/mdeq.nsf/page/Geology_GeologicMapofMississippiShapeFiles?OpenDocument.

Mississippi Department of Environmental Quality. 1989. *The Value of Geologic Mapping in Mississippi*. Bureau of Geology Pamphlet 2.

Mississippi State Oil and Gas Board. 2015. "Annual Production." http://www.ogb.state.ms.us/annprod.php.

Mississippi State University. 2015. *2015 Mississippi Agriculture, Forestry and Natural Resources*. Division of Agriculture, Forestry, and Veterinary Medicine. Available at http://www.dafvm.msstate.edu/factbook.pdf.

Mitchell, W. A, and C. J. Newling. 1986. *Greentree Reservoirs: Section 5.5.3, US Army Corps of Engineers, Wildlife Resources Management Manual*. Technical Report EL-86-9.

Moore, W. H. 1974. *Tinsley Field 1939–1974: A Commemorative Bulletin*. Mississippi Geological, Economic, and Topographic Survey Bulletin 119. Available at https://www.deq.state.ms.us/mdeq.nsf/page/Geology_Bulletin119 TINSLYFIELD1939_1974ACOMMEMORATIVEBULLETIN?Open Document.

Morton, R. A. 2008. "Historical Changes in the Mississippi-Alabama Barrier-Island Chain and the Roles of Extreme Storms, Sea Level, and Human Activities." *Journal of Coastal Research* 24 (6): 1587–1600.

National Oceanic and Atmospheric Administration. 2005. *Hurricane Katrina*. National Climatic Data Center. Available at http://www.ncdc.noaa.gov/extremeevents/specialreports/Hurricane-Katrina.pdf.

National Oceanic and Atmospheric Administration. n.d. "Hurricane Katrina 2005." http://www.nhc.noaa.gov/outreach/history/#katrina.

National Oceanic and Atmospheric Administration. n.d. Mississippi Sound, MS/LA/AL (G170). http://estuarinebathymetry.noaa.gov/bathy_htmls /G170.html.

National Weather Service Forecast Office. 2015. "Hurricane Katrina—A Look Back 10 Years Later." http://www.srh.noaa.gov/lix/?n=katrina_anniversary.

Nature Conservancy. n.d. "Mississippi: Pascagoula River Watershed." http://www.nature.org/ourinitiatives/regions/northamerica/unitedstates /mississippi/placesweprotect/pascagoula-river-watershed.xml.

Nowell, P. W. 2006. "The Flood of 1927 and its Impact in Greenville, Mississippi." Mississippi History Now. http://mshistorynow.mdah.state.ms.us /articles/230/the-flood-of-1927-and-its-impact-in-greenville-mississippi.

Nuwer, D. S. 2005. "Gambling in Mississippi: Its Early History." Mississippi History Now. http://mshistorynow.mdah.state.ms.us/articles/80/gambling-in-mississippi-its-early-history.

Oivanki, S. M. 2005. "Beach and Nearshore Sediment Budget of Harrison County, Mississippi: A Historical Analysis." In *Mississippi Coastal Geology and Regional Marine Study 1990–1994*, vol. 1, 437–489. St. Petersburg, FL: US Geological Survey Center for Coastal Geology and Regional Marine Studies. Available at http://geology.deq.state.ms.us/coastal/NOAA_DATA /Publications/Publications/Coastwide/Volume%201%20Complete.pdf.

Oivanki, S. M. 2005. "Belle Fontaine Project." In *Mississippi Coastal Geology and Regional Marine Study 1990–1994*, vol. 4, 537–541. St. Petersburg, FL: US Geological Survey Center for Coastal Geology and Regional Marine Studies. Available at http://geology.deq.state.ms.us/coastal/NOAA_DATA /Publications/Publications/Coastwide/volume%204.pdf.

Oivanki, S. M. 2005. "Hancock County Beach Project." *Mississippi Coastal Geology and Regional Marine Study, 1990–1994*, vol. 4, 490–519. St. Petersburg, FL: US Geological Survey Center for Coastal Geology and Regional Marine Studies. Available at http://geology.deq.state.ms.us/coastal/NOAA_DATA /Publications/Publications/Coastwide/volume%204.pdf.

Oivanki, S. M. 2005. "Mississippi Mainland Shoreline Beach and Nearshore Profiles." In *Mississippi Coastal Geology and Regional Marine Study, 1990–1994*, vol. 4, 411–436. St. Petersburg, FL: US Geological Survey Center for Coastal Geology and Regional Marine Studies. Available at http://geology .deq.state.ms.us/coastal/NOAA_DATA/Publications/Publications/Coast wide/volume%204.pdf.

Oivanki, S. M. 2005. "Mississippi Shoreline Geomorphology." In *Mississippi Coastal Geology and Regional Marine Study 1990–1994*, vol. 1, 5–26. St. Petersburg, FL: US Geological Survey Center for Coastal Geology and Regional Marine Studies. Available at http://geology.deq.state.ms.us /coastal/NOAA_DATA/Publications/Publications/Coastwide/Volume%20 1%20Complete.pdf.

Oivanki, S. M. 2005. "Round Island Project." In *Mississippi Coastal Geology and Regional Marine Study 1990–1994*, vol. 4, 520–536. St. Petersburg, FL: US Geological Survey Center for Coastal Geology and Regional Marine Studies. Available at http://geology.deq.state.ms.us/coastal/NOAA_DATA/Publications/Publications/Coastwide/volume%204.pdf.

Oivanki, S. M. 2005. "Summary and Conclusions." In *Mississippi Coastal Geology and Regional Marine Study 1990–1994*, vol. 4, 542–551. St. Petersburg, FL: US Geological Survey Center for Coastal Geology and Regional Marine Studies. Available at http://geology.deq.state.ms.us/coastal/NOAA_DATA/Publications/Publications/Coastwide/volume%204.pdf.

Otvos, E. G. 1987. "Late Neogene Stratigraphic Problems in Coastal Mississippi and Alabama." *Mississippi Geology* 7 (3): 8–12.

Otvos, E. G. 1994. "Mississippi's Revised Neogene Stratigraphy in the Northern Gulf Context." In *Gulf Coast Association Geological Societies Transactions*, vol. 44, 541–554.

Otvos, E. G. 1997. *Northeastern Gulf Coastal Plain Revisited: Neogene and Quaternary Units and Events: Old and New Concepts.* Guidebook, Gulf Coast Association of Geological Societies. New Orleans: New Orleans Geological Society.

Otvos, E. G. 1998. "The Gulf Coastal Citronelle Formation: Definition, Depositional Facies, and Age Issue." In *Gulf Coast Association of Geological Societies Transactions*, vol. 48, 321–333.

Otvos, E. G. 2001. *Mississippi Coast: Stratigraphy and Quaternary Evolution in the Northern Gulf Coastal Plain Framework.* US Geological Survey Open-File Report 01-415-H. Available at https://pubs.usgs.gov/of/2001/of01-415/chap8txt.pdf.

Otvos, E. G. 2005. "Numerical Chronology of the Late Quaternary Gulf Coastal Plain, Barrier Evolution and an Updated Holocene Sea-Level Curve." In *Gulf Coast Association of Geological Societies Transactions*, vol. 55, 629–641.

Otvos, E. G. 2015. "The Last Interglacial Stage: Definitions and Marine Highstand, North America and Eurasia." *Quaternary International* 383: 158–173.

Otvos, E. G., and G. A. Carter. 2013. "Regressive and Transgressive Barrier Islands on the North-Central Gulf Coast: Contrasts in Evolution, Sediment Delivery, and Island Vulnerability." *Geomorphology* 198: 1–19.

Otvos, E. G., and M. J. Giardino. 2004. "Interlinked Barrier Chain and Delta Lobe Development, Northern Gulf of Mexico." *Sedimentary Geology* 169 (1–2): 47–73.

Parks, W. S., W. H. Moore, T. E. McCutcheon, and B. E. Wasson. 1963. *Attala County Surface Geology.* Mississippi Geological, Economic, and Topographical Survey Bulletin 99. Available at https://www.deq.state.ms.us/MDEQ.nsf/page/Geology_Bulletin99AttalaCountyMineralResources?OpenDocument.

Pashin, J. C. 1994. "Cycles and Stacking Patterns in Carboniferous Rocks of the Black Warrior Foreland Basin." In *Gulf Coast Association of Geological Societies Transactions*, vol. 44, 555–563.

PBS. n.d. American Experience. "The Mounds Levee Landing Break." http://www.pbs.org/wgbh/americanexperience/features/general-article/flood-levee/.

Pirkle, F. L., W. A. Pirkle, and E. C. Pirkle. 2007. "Heavy-Mineral Sands of the Atlantic and Gulf Coastal Plains." In *Heavy Minerals in Use*, Developments in Sedimentology 58, edited by M. A. Mange and D. T. Wright, 1144–1232. Oxford: Elsevier.

Priddy, R. R. 1961. *Geologic Study Along Highway 80 from Alabama Line to Jackson, Mississippi*. Mississippi Geological Survey Bulletin 91. Available at http://deq.state.ms.us/mdeq.nsf/page/Geology_Bulletin91GeologicStudy AlongHighway80FromAlabamaLinetoJacksonMississippi.

Puckett, T. M. 1991. "Absolute Paleobathymetry of Upper Cretaceous Chalks Based on Ostracodes—Evidence from the Demopolis Chalk (Campanian and Maastrichtian) of the Northern Gulf Coastal Plain." *Geology* 19 (5): 449–452.

"Q&A: Texas Supercollider Project Scrapped." *Tampa Bay Times* (December 30, 2009). http://www.tampabay.com/news/humaninterest/qampa-texas-supercollider-project-scrapped/1062063.

Renken, R. A. 1998. *Ground Water Atlas of the United States: Arkansas, Louisiana, Mississippi*. United States Geological Survey HA 730-F. Available at http://pubs.usgs.gov/ha/ha730/ch_f/.

Rittenour, T. M., M. D. Blum, and R. J. Goble. 2007. "Fluvial Evolution of the Lower Mississippi River Valley During the Last 100 K. Y. Glacial Cycle: Response to Glaciation and Sea-Level Change." *Geological Society of America Bulletin* 119 (5–6): 586–608.

Robertson, P. B., and M. D. Butler. 1982. "New Evidence for the Impact Origin of Kilmichael Mississippi." *Lunar and Planetary Science XIII*, 653–654. Houston: The Institute.

Rogers, J. D. n.d. "Evolution of the Levee System along the Lower Mississippi River." Natural Hazards Mitigation Institute, Department of Geological Engineering, University of Missouri—Rolla. http://web.mst.edu/~rogersda /levees/Evolution%20of%20the%20Levee%20System%20Along%20the%20 Mississippi.pdf.

Russell, E. E. 1986. "Shelf Marls and Chalks in the Marine Section of the Upper Cretaceous: Mississippi." In *Southeastern Section of the Geological Society of America*, Centennial Field Guide, vol. 6, edited by T. L. Neathery, 387–392. Boulder, CO: Geological Society of America.

Rutledge, E. M., M. J. Guccione, H. W. Markewich, D. A. Wysocki, and L. B. Ward. 1996. "Loess Stratigraphy of the Lower Mississippi Valley." *Engineering Geology* 45 (1–4): 167–183.

Salvador, A. 1991. "Origin and Development of the Gulf of Mexico Basin." In *The Geology of North America*, vol. J, edited by A. Salvador. Boulder, CO: Geological Society of America.

Sanders, P. 2006. "Casinos Emerge as Winners in Wake of Hurricane Katrina." *Wall Street Journal* (August 3). http://www.wsj.com/articles/SB115456892666425327.

Saucier, R. T. 1994. *Geomorphology and Quaternary Geologic History of the Lower Mississippi Valley*. Vols. I and II. Vicksburg, MS: US Army Engineer Waterways Experiment Station, Corps of Engineers.

Saucier, R. T., and C. R. Kolb. 1967. *Alluvial Geology of the Yazoo Basin, Lower Mississippi Valley*. Vicksburg, MS: US Army Engineer Waterways Experiment Station, Corps of Engineers.

Saunders, J. A. 1992. "Age and Petrology of the Jackson Dome Igneous-Volcanic Complex, Mississippi: Implications for the Tectonic History of the Mississippi Salt Dome Basin." In *Gulf Coast Association of Geological Societies Transactions*, vol. 42, 659–667.

Schmid, K. 2000. *Historical Evolution of Mississippi's Barrier Islands*. Mississippi Department of Environmental Quality, Office of Geology Report. Available at http://geology.deq.state.ms.us/coastal/NOAA_DATA/Publications/Publications/Barrier_Islands/Historical%20Evolution%20of%20Mississippi%20Barrier%20Islands.pdf.

Schmid, K. 2001. *Using Vibracore and Profile Data to Quantify Volumes of Renourished Sediments, Holocene Thickness, and Sedimentation Patterns: Hancock County, Mississippi*. Mississippi Department of Environmental Quality, Office of Geology Open File Report 131.

Seal, M. C. 1985. *Stratigraphic-Structural Cross Section, Desoto County, Mississippi*. Mississippi Bureau of Geology. Not published.

Smith, L. M. 1996. "Fluvial Geomorphic Features of the Lower Mississippi Alluvial Valley." *Engineering Geology* 45 (1–4): 139–165.

Smith, M. L., and M. A. Meylan. 1983. "Red Bluff, Marion County, Mississippi: A Citronelle Braided Stream Deposit." In *Gulf Coast Association of Geological Societies Transactions*, vol. 33, 419–432.

Snowden, J. O., R. R. Priddy, and C. D. Caplenor. 1968. *Loess Investigations in Mississippi*. Mississippi Geological, Economic, and Topographical Survey Bulletin 111. Available at https:/www.deq.state.ms.us/MDEQ.nsf/page/Geology_Bulletin111?OpenDocument.

Starnes, J. E. 2009. "Precious Opal: Mississippi's First Gemstone." *Mississippi Department of Environmental Quality, Environmental News* 6 (12): 82. Available at http://www.deq.state.ms.us/MDEQ.nsf/pdf/Geology_JustGeology20082009/$File/89850%20Just%20Geology%20Book%20Final.pdf?OpenElement.

Sundeen, D. A., and P. L. Cook. 1977. "K-Ar Dates from Upper Cretaceous Volcanic Rocks in the Subsurface of West-Central Mississippi." *Geological Society of America Bulletin* 88 (8): 1144–1146.

Taylor, R. E. 1972. *Geohydrology of Tatum Salt Dome Area, Lamar and Marion Counties, Mississippi.* United States Geological Survey Report VUF-1023.

Tennessee-Tombigbee Waterway. n.d. "History of the Tenn-Tom." http://history.tenntom.org/.

Texas A & M University. n.d. Bathymetry and Coastlines for the Gulf of Mexico. http://gcoos.tamu.edu/products/topography/Introduction.html.

Thomas, W. A. 1977. "Evolution of Appalachian-Ouachita Salients and Recesses from Reentrants and Promontories in the Continental Margin." *American Journal of Science* 277 (December): 1233–1278.

Thomas, W. A. 1988. "The Black Warrior Basin." In *Sedimentary Cover—North American Craton: US*, the Geology of North America, vol. D-2, edited by L. L. Sloss, 471–492. Boulder, CO: Geological Society of America.

Thomas, W. A. 1991. "The Appalachian-Ouachita Rifted Margin of Southeastern North America." *Geological Society of America Bulletin* 103: 415–431.

Thomas, W. A. 1993. "Low-Angle Detachment Geometry of the Late Precambrian-Cambrian Appalachian-Ouachita Rifted Margin of Southeastern North America." *Geology* 21 (10): 921–924.

Thomas, W. A. 2004. "Genetic Relationship of Rift-Stage Crustal Structure, Terrane Accretion, and Foreland Tectonics Along the Southern Appalachian-Ouachita Orogen." *Journal of Geodynamics* 37 (3–5): 549–563.

Thompson, D. E. 2009. *Structural Features of Mississippi.* Mississippi Department of Environmental Quality, Office of Geology. Available at http://www.deq.state.ms.us/mdeq.nsf/pdf/Geology_StructureFeatMS072009/$File/Struc_Feat_MS.pdf?OpenElement.

Thompson, D. E. 2011. *Geologic Map of Mississippi.* Mississippi Department of Environmental Quality, Office of Geology. Available at http://www.deq.state.ms.us/mdeq.nsf/pdf/Geology_MSGeology1969Map/$File/MS_Geology1969.pdf?OpenElement.

Thompson, D. E. 2014. *Geologic Map of the Holly Springs 1:100,000 Quadrangle.* Mississippi Department of Environmental Quality, Office of Geology Open-File Report OF-276. Available at http://www.deq.state.ms.us/mdeq.nsf/pdf/Geology-GeologicMapHollySprings/$File/HollSprings_1_100000.pdf?OpenElement.

Unisys Weather. 2015. Atlantic Tropical Storm Tracking by Year. http://weather.unisys.com/hurricane/atlantic/.

US Army Corps of Engineers. n.d.(a). *Geological Highlights of the Vicksburg Area.* Engineering Geology and Rock Mechanics Division, US Army Corps of Engineers Waterways Experiment Station, Geotechnical Laboratory.

US Army Corps of Engineers. n.d.(b). "Grenada Lake History." http://www .mvk.usace.army.mil/Missions/Recreation/GrenadaLake/Historyand Mission.aspx.

US Army Corp of Engineers. n.d(c). "Mississippi Rivers and Tributaries Project." http://www.mvd.usace.army.mil/About/Mississippi-River-Commis sion-MRC/Mississippi-River-Tributaries-Project-MR-T/.

US Department of Energy. 2012. *Salmon, Mississippi, Site.* Fact Sheet. Available at http://www.lm.doe.gov/salmon/Documents.aspx.

US Geological Survey. 1988. *Phoenix, Mississippi, 7.5 Minute Topographic Map.*

US Geological Survey Geologic Names Committee. 2010. *Divisions of Geologic Time—Major Chronostratigraphic and Geochronologic Units.* US Geological Survey Fact Sheet 2010–3059. Available at http://pubs.usgs.gov/ fs/2010/3059/.

Van Arsdale, R. B., and R. K. TenBrink. 2000. "Late Cretaceous and Cenozoic Geology of the New Madrid Seismic Zone." *Bulletin of the Seismological Society of America* 90 (2): 345–356.

Waldron, B., D. Larsen, R. Hannigan, R. Csontos, J. Anderson, C. Dowling, and J. Bouldin. 2011. *Mississippi Embayment Regional Ground Water Study.* EPA 600/R-10/130. Office of Research and Development, National Risk Management Research Laboratory.

Walkinshaw, S. 2008. *North-South Cross Section Through the Jackson Dome.* Madison, MS: Vision Exploration. Personal communication.

Walper, J. L. 1974. "The Origin of the Bahama Platform." *Gulf Coast Association of Geological Societies Transactions*, vol. 24, 25–30.

Walper, J. L., and C. L. Rowett. 1972. "Plate Tectonics and the Origin of the Caribbean Sea and the Gulf of Mexico." *Gulf Coast Association of Geological Societies Transactions*, vol. 22, 105–116.

Wascher, H. L., R. P. Humbert, and J. G. Cady. 1948. "Loess in the Southern Mississippi Valley: Identification and Distribution of the Loess Sheets." *Soil Science Society of America Journal* 12: 389–399.

Williams, C. H. 1969. *Cross Section from Mississippi-Tennessee State Line to Horn Island in the Gulf of Mexico.* Mississippi Geological Survey CS-1. Available at https://www.deq.state.ms.us/MDEQ.nsf/page/Geology_cross _sections_publications?OpenDocument.

Willyard, C. 2008. "Desperately Seeking Salt Dome." Geotimes (July). http: //www.geotimes.org/july08/article.html?id=nn_salt.html.

Wilson, G. V. 1975. "Early Differential Subsidence and Configuration of the Northern Gulf Coast Basin in Southwest Alabama and Northwest Florida." In *Gulf Coast Association of Geological Societies Transactions*, vol. 25, 196–206.

INDEX

Stan Galicki has thirty-three years of experience as a geologist, twenty-four as a professor at Millsaps College in Jackson, Mississippi. He worked in petroleum exploration prior to taking on academic responsibilities. His primary research fields include sedimentary depositional environments, wetland biogeochemistry, and dendrochronology. Stan is a registered professional geologist and member of the Association of Environmental and Engineering Geologists and the Geological Society of America.

Darrel Schmitz has thirty-five years of experience as a geologist and has spent the last twenty-five as a professor at Mississippi State University. He is a registered professional geologist in Mississippi and has experience primarily with the development and protection of water and fossil fuel resources. Darrel is active in professional organizations, having served as president of both the National Association of State Boards of Geology and the Association of Environmental and Engineering Geologists. He is also a fellow of the Geological Society of America.